SpringerBriefs in Operations Management

Series Editor

Suresh P. Sethi
The University of Texas at Dallas, TX, USA

More information about this series at http://www.springer.com/series/13082

Xiaoying Liang · Lijun Ma
Haifeng Wang · Houmin Yan

Inventory Management with Alternative Delivery Times

Springer

Xiaoying Liang
College of Business
City University of Hong Kong
Kowloon
Hong Kong

Lijun Ma
College of Management
Shenzhen University
Shenzhen, Guangdong
China

Haifeng Wang
Fengyuan Innovation Technology Co., Ltd.
Beijing
China

Houmin Yan
College of Business
City University of Hong Kong
Kowloon
Hong Kong

ISSN 2365-8320 ISSN 2365-8339 (electronic)
SpringerBriefs in Operations Management
ISBN 978-3-319-48633-8 ISBN 978-3-319-48635-2 (eBook)
DOI 10.1007/978-3-319-48635-2

Library of Congress Control Number: 2016955915

Printed on acid-free paper

This Springer imprint is published by Springer Nature
The registered company is Springer International Publishing AG
The registered company address is: Gewerbestrasse 11, 6330 Cham, Switzerland

To our families and friends

Preface

Inventory management constitutes an important research area in operations management. The research has mainly focused on the problem of how inventory managers should make their procurement decisions and allocate limited inventory to satisfy customers' demand. Around this central problem, a wide variety of interesting topics have been examined, including the effect of the procurement cost structure on inventory policies, demand learning and forecasting, and lead-time management.

Previous work on lead-time management has primarily focused on the supply side. Supply-side lead-time corresponds to the delay between the time an inventory manager places a replenishment order and the time the replenishment actually arrives. The problem has been considered in a number of contexts, including stochastic lead times, multiple supply sources with different procurement costs and lead times, and multi-echelon, to name a few. There has also recently been an increasing interest in applying the idea of lead-time management to the demand side. Demand-side lead-time can be interpreted as the delivery delay between the time a customer places an order and the time the order arrives. Demand-side lead-time has gained attention in part because firms have begun to realize that customers have heterogeneous delivery-time requirements, and they can exploit this heterogeneity to their advantage by offering multiple delivery-time options to customers. This delivery-time-based differentiation strategy is particularly useful when firms have scarce inventory and face a long replenishment lead-time on the supply side. Although industry practices of such a strategy are abundant, a more rigorous theoretical treatment is still lacking and would be of great practical value to an effective implementation of the strategy.

We develop a modeling framework to analyze the problem of inventory management with alternative delivery times in this book. The general context considered here is that a seller replenishes its inventory in fixed intervals and, between replenishments, allocates the limited inventory to satisfy customers who are both price and delivery-time sensitive. On the demand side, customers have heterogeneous delivery-time requirements and choose either spot or late delivery. This theoretical modeling captures the essence of real-world business practices such as the delivery-time market segmentation strategy adopted by automobile dealerships in China and many other similar examples. We focus on the seller's optimal inventory replenishment and demand fulfillment policies, and our results provide managerial insights into the merits of flexible delivery-time options. Similar applications such as the group-buying mechanism are also examined. The main mathematical tool used in theoretical analysis is dynamic programming. This book is written for students, researchers, and practitioners in the areas of operations management and industrial engineering who are interested in understanding the rationale of flexible delivery times and designing successful applications.

We wish to thank Zhiyuan Chen, Meng Lu, Suresh Sethi, Lei Xie, and Yangyang Xie, who have worked with us in the area of inventory management with alternative delivery times. This book is supported in part by GRF Grants 9041969, RGC Theme-based Research Scheme T32-620/11, City University of Hong Kong Startup Grant 7200290, NSFC Grants 71001073, 71271182, 71302189, 71390335 and 71471118, the Humanities and Social Sciences Foundation of Ministry of Education of China (No. 14YJC630096), and the Distinguished University Young Scholar Program of Guangdong Province (No. Yq2013140). We are grateful for their financial support.

Hong Kong, China — Xiaoying Liang
Shenzhen, China — Lijun Ma
Beijing, China — Haifeng Wang
Hong Kong, China — Houmin Yan
August 2016

Contents

Chapter 1
Overview

Abstract Firms now recognize the important role of delivery time in both customer segmentation and inventory management. In this chapter, we discuss the merits of delivery-time management and its practical implications. We then highlight our research presented in this book. Related literature is reviewed at the end of this chapter.

1.1 Introduction

By definition, lead time corresponds to the latency between the initiation and completion of an action or a process. In a supply-chain context, lead time is essential and visible almost everywhere. It is fair to say that where there is an order, there could exist a lead time. Effective lead-time management is thus crucial to the success of supply chain management.

Perhaps the most obvious lead time in a supply chain is the replenishment lead time, which is the time between when a replenishment order is placed and when the order arrives. The role and optimization of the replenishment lead time have been given much attention in both theory and practice. Researchers have previously studied the optimal replenishment policy in the presence of stochastic lead times (Kaplan 1970), multiple delivery modes (Neuts 1964; Lawson and Porteus 2000) and etc. Meanwhile, real-world firms have long recognized that a reduction in the replenishment lead time can significantly improve the responsiveness of a supply chain. For example, H&M and Zara are fast fashion retailers. In addition to their talented instincts about fashion trends and quick response programs with shorter operational cycles, their acclaimed achievements in cutting lead times have led to their phenomenal success (Caro and Martínez-de Albéniz 2010).

What seems less obvious is the lead time on the demand side, i.e., the delay in the fulfillment of customer orders, or simply the delivery time. There are several reasons why delivery time has attracted relatively less attention. First, delivery time is less of a concern for traditional brick-and-mortar stores, in which customers are mostly

X. Liang et al., *Inventory Management with Alternative Delivery Times*,
SpringerBriefs in Operations Management, DOI 10.1007/978-3-319-48635-2_1

fulfilled on the spot from on-hand inventories and the demand is relatively stable and predictable. Second, customer orders are likely to arrive sequentially rather than in a batch as in the case of replenishment orders. Hence, although firms recognize the importance of delivery time, it is hard for them to keep track of, let alone manage, the delivery times of individual customers.

In recent years, the prosperity of e-commerce has given delivery time an increasingly important role. Most online orders are shipped by couriers rather than picked up by customers themselves. Also, with the aid of advances in information technology and logistics, firms are now able to implement a flexible delivery-time management system and offer customers multiple delivery-time options to segment them. In addition to a refined market segmentation, delivery-time management on the demand side has other benefits. For example, delivery time can be used to counteract the supply-side lead time, and information from advance orders can help firms to improve their decision making (Hariharan and Zipkin 1995). Firms may also resort to flexible delivery-time options to cope with unexpected situations such as a severe supply-demand imbalance (Liang et al. 2013).

All of these benefits has made flexible delivery times a popular topic in academia and industry. Specifically, an effective implementation of the strategy depends on details such as the following:

- the commitment control of each incoming customer order;
- the clearance of backlogged orders;
- the replenishment decision in the presence of delivery-time management; and
- joint pricing and delivery decisions.

The discussion of these issues calls for a comprehensive analytical framework. In the last decade, we have been working on the problem of flexible delivery-time management involving topics such as optimal inventory commitment and replenishment policies, joint pricing and inventory-control decisions, and prioritized backlogging clearance, to name just a few. We have formulated real-world problems using mathematical models, develop approaches for their analysis, and interpret theoretical results into managerial insights for the better design and execution of flexible delivery-time management. In this book, we intend to provide a unified treatment of the preceding models, summarize and provide a critique of the major results, and shed light on potential research directions.

1.2 The Aim of This Book

In this book, we follow a progressive approach to lay out our research. In Chap. 2, we introduce in detail the real-world examples that inspire our research. These examples explain the rationale behind firms' applications of flexible delivery times. We illustrate that firms may resort to delivery-time management for a variety of reasons. For example, automobile dealerships offer multiple delivery options to customers when they face a severe supply-demand imbalance. Manufacturers can benefit from flexible delivery times because of the improvement in production planning. Online retailers allow customers to choose from a menu of shipping options for a more effective use of shipping capacities. Service providers make use of alternative delivery times to better segment the market.

We start with a baseline model in Chap. 3 to capture the essence of the examples (the dealership example in particular) presented in Chap. 2. The basic setting is as follows. We consider an inventory system with alternative lead times. The seller replenishes its inventory in cycles and satisfies its customers using commitment control between replenishments. Customers are classified into two segments according to their price and lead-time sensitivities: short and long lead-time customers. The orders from short lead-time customers need to be satisfied on the spot from the on-hand inventory, and the orders from long lead-time customers can be postponed for a late delivery. In compensation for the late delivery, long lead-time customers pay a lower price than their short lead-time counterparts. With such a model, we are able to derive the optimal inventory replenishment and commitment policies.

The baseline model serves as a foundation for a number of extensions, which we analyze in subsequent chapters. As can be seen in many real-world examples, the delivery schedule is flexible rather than rigid. That is, the seller may choose to deliver a backlogged long lead-time order earlier than promised. We call such an action an upgrade and investigate the upgrade mechanism in Chap. 4. Although the upgrade decision complicates the seller's decision-making process, it improves the service level and saves the seller from carrying too much inventory. In addition to obtaining the optimal replenishment and upgrade policies, we further calibrate the model parameters using the empirical data collected from Toyota dealerships and calculate the optimal policies and several performance metrics accordingly. The results indicate that the optimal upgrade policy can improve both profit and inventory use. In Chap. 5, we further incorporate the pricing decision and study the joint pricing and inventory-control policies. Moreover, we provide a preliminary comparison of the two mechanisms to illustrate their respective advantages and disadvantages. In Chap. 6, we focus on the specific problem of how backlogged orders should be cleared when replenishments arrive. We assume that the seller handles backlogged orders with three common priority rules. We characterize, evaluate, and compare the inventory commitment policies under these three priority rules.

Although the setting of the baseline model and its various extensions may appear restrictive, the modeling and analysis approaches are general and can be applied to problems of a similar nature. In Chap. 7, we illustrate this point using the case of group buying. The seller offers a regular spot-selling option and a group-buying option. We show that the same kind of customer segmentation and demand rationing used in the models with alternative delivery lead times also applies here. We characterize the optimal rationing policy and further discuss several extensions. The results are of practical value to firms that offer multiple selling options.

The rest of this chapter reviews the related literature.

1.3 Related Literature

In this section, we review the literature related to our work. We focus on two major streams of research: inventory models with supply-side lead times and inventory systems serving multiple demand classes.

There is a large body of literature on the optimal inventory policy for either deterministic or stochastic supply-side lead time. Fukuda (1964) studies a multi-period inventory system with two deterministic lead-time options and characterizes the optimal threshold policies in both cases with and without the setup cost. Neuts (1964) and Lawson and Porteus (2000) extend Fukuda (1964) to further consider multiple delivery modes. Kaplan (1970) discusses a dynamic inventory problem with random delivery lead times, in which a no-crossover assumption is imposed. He shows that the optimal policies are similar to those obtained in the case of deterministic lead times. Ehrhardt (1984) solves an infinite-horizon model with stochastic lead times and obtains the optimal policy under both discounted- and average-cost criteria. Song and Zipkin (1996) investigate an inventory system in which the replenishment lead time follows a Markov process. Song (1994a, b) elaborates on the effect of lead-time uncertainty on inventory policy and optimal cost using measures of long-run average cost and discounted cost, respectively. Chen (2001), Ozer (2003), and Ozer and Wei (2006) study inventory replenishment and pricing issues in periodic-review inventory systems in which the demand is segmented by delivery lead-time requirements. Lederer and Li (1997) look at the competition between firms serving lead-time-sensitive customers. As mentioned earlier, our study focuses on the role of demand-side delivery lead times. On the supply side, the optimal inventory replenishment policies resemble those in the literature. On the demand side, firms can use the heterogeneity in customers' delivery-time requirement to their advantage. For example, when the on-hand inventory is running low, firms can backlog patient customers for late fulfillment and reserve on-hand inventory for impatient yet high-margin customers. When the on-hand inventory is abundant, firms can choose to fill all customer orders on the spot to reduce inventory carrying costs.

When there are multiple demand classes and their demands exceed the limited inventory, a multi-level rationing policy is often used to achieve the optimal profit by reserving different levels of inventory for different demand classes. There are three common modeling frameworks: periodic-review, continuous-review, and queuing inventory models. In the context of periodic-review inventory models, the pioneering work of Veinott (1965) introduces the concept of rationing to satisfy the demands from multiple demand classes. Topkis (1968) further refines the model by dividing the replenishment cycle into a finite number of intervals. The on-hand inventory is allocated in each interval, and unsatisfied demand can be either backlogged or lost. Kaplan (1969) independently obtains the same inventory rationing policy as that in Topkis (1968) for the backlog case. Cattani and Souza (2002) study an inventory system serving customers with two lead-time requirements and compare numerically the performance of the inventory rationing policy and the simple priority rules policy. Jang (2006) and Jang and Kim (2007) characterize the production, inventory rationing, and distribution decisions under a newsvendor-type model in which there are two classes of customers with different waiting costs. Frank et al. (2003) investigate a periodic-review inventory system serving a deterministic and a stochastic demand streams. The deterministic demand must be satisfied, and the inventory replenishment incurs a setup cost. An inventory rationing policy is used to reserve inventory for the committed deterministic demand. Ding et al. (2006) develop an inventory rationing model that incorporates a price-discount mechanism and allows backlogging. Duran et al. (2008) discuss a multi-period inventory system where demands are differentiated according to their lead-time requirements, designated as Shorts and Longs. At the end of each period, after the Shorts and Longs materialize, the supplier decides the amount of inventory to reserve, the amount of Longs to backlog, and the amount to replenish. They obtain the optimal (S, R, B) policy, where S, R, and B represent the order-up-to, reserve-up-to, and backlog-up-to levels, respectively. Gupta and Wang (2007) examine a periodic-review infinite-horizon inventory model serving both transactional and contractual customers in the make-to-order and make-to-stock/make-to-order environments. They prove that a threshold policy is optimal for the make-to-order mode and that a two-critical-number policy performs well for the make-to-stock/make-to-order mode.

In the context of continuous-review inventory models, Deshpande et al. (2003) analyze the performance of a static threshold rationing policy, under the assumption that the inventory is replenished according to a (Q, r) policy (i.e., when the inventory position falls to the level r, a replenishment order of Q units is placed). Arslan et al. (2007) consider a model with deterministic replenishment lead time using the (Q, r) policy to fulfill customers with different shortage costs. They derive efficient heuristic algorithms to ration inventory for different customer classes.

Queuing models are also popular in the study of the optimal production control and inventory rationing policies. Ha (1997a, b) analyze a single-product production system that operates in a make-to-stock environment with backorders and lost sales. De Véricourt et al. (2001, 2002) extend Ha (1997b) to a manufacturing system with multi-class customers. Carr and Duenyas (2000) extend these models further to a multi-product manufacturing system to address the sequencing issues under admission control. Chen et al. (2007) study the optimal admission control of selling channels for an etailer. The etailer can sell products on the websites of one or more third parties under the cost-per-click scheme.

We use a periodic-review setting throughout our analysis. Compared with traditional periodic-review inventory rationing models, we assume that the seller allocates goods upon the arrival of a customer (in real time) rather than dispatches goods to customers when all demands are realized. This is of practical interest, as Ding et al. (2006) point out, because "in most practical cases, demand arrives according to a continuous-time stochastic process during a period and firms are required to respond to demands immediately." In our models, the inventory dynamics depend on the arrival sequence of both short and long lead-time customers and the admission control deployed.

References

H. Arslan, S.C. Graves, T.A. Roemer, A single-product inventory model for multiple demand classes. Manag. Sci. **53**(9), 1486–1500 (2007)

F. Caro, V. Martínez-de, Albéniz, The impact of quick response in inventory-based competition. Manuf. Serv. Oper. Manag. **12**(3), 409–429 (2010)

S. Carr, I. Duenyas, Optimal admission control and sequencing in a make-to-stock/make-to-order production system. Oper. Res. **48**(5), 709–720 (2000)

K.D. Cattani, G.C. Souza, Inventory rationing and shipment flexibility alternatives for direct market firms. Prod. Oper. Manag. **11**(4), 441–457 (2002)

F. Chen, Market segmentation, advanced demand information, and supply chain performance. Manuf. Serv. Oper. Manag. **3**(1), 53–67 (2001)

F.Y. Chen, J. Chen, Y.B. Xiao, Optimal control of selling channels for an online retailer with cost-per-click payments and seasonal products. Prod. Oper. Manag. **16**(3), 292–305 (2007)

F. De Véricourt, F. Karaesmen, Y. Dallery, Assessing the benefits of different stock-allocation policies for a make-to-stock production system. Manuf. Serv. Oper. Manag. **3**(2), 105–121 (2001)

F. De Véricourt, F. Karaesmen, Y. Dallery, Optimal stock allocation for a capacitated supply system. Manag. Sci. **48**(11), 1486–1501 (2002)

V. Deshpande, M.A. Cohen, K. Donohue, A threshold inventory rationing policy for service-differentiated demand classes. Manag. Sci. **49**(6), 683–703 (2003)

Q. Ding, P. Kouvelis, J.M. Milner, Dynamic pricing through discounts for optimizing multiple-class demand fulfillment. Oper. Res. **54**(1), 169–183 (2006)

S. Duran, T. Liu, D. Simchi-Levi, J.L. Swann, Policies utilizing tactical inventory for service-differentiated customers. Oper. Res. Lett. **36**(2), 259–264 (2008)

R. Ehrhardt, (s, S) policies for a dynamic inventory model with stochastic lead times. Oper. Res. **32**(1), 121–132 (1984)

K.C. Frank, R.Q. Zhang, I. Duenyas, Optimal policies for inventory systems with priority demand classes. Oper. Res. **51**(6), 993–1002 (2003)

Y. Fukuda, Optimal policies for the inventory problem with negotiable leadtime. Manag. Sci. **10**(4), 690–708 (1964)

D. Gupta, L. Wang, Capacity management for contract manufacturing. Oper. Res. **55**(2), 367–377 (2007)

A.Y. Ha, Inventory rationing in a make-to-stock production system with several demand classes and lost sales. Manag. Sci. **43**(8), 1093–1103 (1997a)

A.Y. Ha, Stock-rationing policy for a make-to-stock production system with two priority classes and backordering. Nav. Res. Logist. **44**(5), 457–472 (1997b)

R. Hariharan, P. Zipkin, Customer-order information, leadtimes, and inventories. Manag. Sci. **41**(10), 1599–1607 (1995)

W. Jang, Production and allocation policies in a two-class inventory system with time and quantity dependent waiting costs. Comput. Oper. Res. **33**(8), 2301–2321 (2006)

W. Jang, D. Kim, Integrated production and allocation policies with one direct shipping option. Eur. J. of Oper. Res. **181**(2), 716–732 (2007)

A. Kaplan, Stock rationing. Manag. Sci. **15**(5), 260–267 (1969)

R.S. Kaplan, A dynamic inventory model with stochastic lead times. Manag. Sci. **16**(7), 491–507 (1970)

D.G. Lawson, E.L. Porteus, Multistage inventory management with expediting. Oper. Res. **48**(6), 878–893 (2000)

P.J. Lederer, L. Li, Pricing, production, scheduling, and delivery-time competition. Oper. Res. **45**(3), 407–420 (1997)

X. Liang, L. Ma, H. Yan, Toyota china: matching supply with demand, in *The Supply Chain Management Casebook: comprehensive Coverage and Best Practices in SCM*, ed. by C. Munson (Financial Times Press, New Jersey, 2013), pp. 68–79

M.F. Neuts, An inventory model with an optional time lag. J. of the Soc. for Ind. Appl. Math. **12**(1), 179–185 (1964)

O. Ozer, Replenishment strategies for distribution systems under advance demand information. Manag. Sci. **49**(3), 255–272 (2003)

O. Ozer, W. Wei, Strategic commitments for an optimal capacity decision under asymmetric forecast information. Manag. Sci. **52**(8), 1238–1257 (2006)

J.S. Song, The effect of leadtime uncertainty in a simple stochastic inventory model. Manag. Sci. **40**(5), 603–613 (1994a)

J.S. Song, Understanding the leadtime effects in stochastic inventory systems with discounted costs. Oper. Res. Lett. **15**, 85–93 (1994b)

J.S. Song, P.H. Zipkin, Inventory control with information about supply conditions. Manag. Sci. **42**(10), 1409–1419 (1996)

D.M. Topkis, Optimal ordering and rationing policies in a nonstationary dynamic inventory model with n demand classes. Manag. Sci. **15**(3), 160–176 (1968)

A.F. Veinott, Optimal policy in a dynamic, single product, nonstationary inventory model with several demand classes. Oper. Res. **13**(5), 761–778 (1965)

Chapter 2
Examples from Industry

Abstract In this chapter, we provide a number of industry examples of how firms implement the strategy of flexible delivery times. The first example is Toyota China's dealerships. Dealerships use a price- and delivery-time-based segmentation strategy to ameliorate the supply-demand imbalance due to inaccurate demand forecast and a rigid production schedule. We also discuss examples such as multiple shipping options offered by online retailers and the priority queue system as commonly seen in the service industry. In all of these examples, firms segment customers based on their price and delivery-time sensitivities, and the critical question is how firms should implement the strategy optimally in a dynamically changing context. This question is elucidated by the theoretical discussion in the following chapters.

2.1 Introduction

Flexible delivery-time options have much practical value and can be widely seen in business practices. Firms make use of these options when they have limited inventories, have trouble in adjusting their production capacities to keep up with surging demand, or simply would like to improve on market segmentation. Delivery-time options segment customers based on their price and delivery-time sensitivities. Firms can thus generate a higher revenue from the on-hand inventory because this specific separating strategy, as noted by Shugan (2002), allows them to "keep price-insensitive customers buying at the *peak* period" for a higher margin and "shift only price-sensitive customers to the *off-peak* period" (peak periods are periods in which the demand far exceeds the capacity, and off-peak periods are defined as the opposite).

This chapter introduces a number of industry examples of the delivery-time segmentation strategy. Section 2.2 presents in detail the practice at Toyota China's dealerships. Dealerships use flexible delivery times to ameliorate the supply-demand imbalance caused by forecasting inaccuracy and rigid production schedules. We use data collected from this dealership example later in our numerical studies. Section 2.3 discusses several other examples, in which flexible delivery times benefit firms in a variety of ways. In Sect. 2.4, we summarize the managerial implications of these examples and conclude the chapter.

X. Liang et al., *Inventory Management with Alternative Delivery Times*,
SpringerBriefs in Operations Management, DOI 10.1007/978-3-319-48635-2_2

2.2 Toyota China's Automobile Dealerships

In this section, we introduce the practice of flexible delivery times adopted by Toyota China's dealerships. In recent years, we have observed that pricing and delivery-time considerations are playing an increasingly important role in the selling of cars in China. In the fast-growing Chinese automobile market, competition between automobile supply chains, consisting mainly of automobile manufacturers and dealerships, has also intensified. In particular, the strategy for automobile supply chains has changed from a simple strategy of access to the market (placement) to a customer-focused strategy (product and price), in which providing a number of different delivery-time options along with differentiated pricing is a tool for maximizing revenue and expanding market share. In the following, we describe how Toyota China's dealerships make use of flexible delivery times to handle the supply-demand imbalance.

2.2.1 Toyota in the Chinese Automobile Market

The automobile industry is one of the most important economic sectors in the world. Based on the total number of vehicles produced in 2005 (66 million), it was estimated that the global turnover of the automobile industry was equivalent to 1.9 trillion euros. The industry also created over 8 million jobs directly, representing more than 5 % of the world's total manufacturing employment, and about five times more jobs indirectly. According to a study of 26 countries conducted by the International Organization of Motor Vehicle Manufacturers (OICA), vehicle manufacturing and usage contributed more than 433 billion euros to government revenues.[1]

The automobile industry in China has a history dating back to the first Five-Year Plan (1953–1957). In 1953, First Automobile Works (FAW) was founded in Changchun, the capital of Jilin province. In the more than five decades since then, and especially since 1978, when China began its economic reform (also known as "reform and opening up"), China's automobile industry has gradually grown into a pillar of the economy and has become one of the most important in the world. Between 2000 and 2011, it recorded an average annual growth rate of 22 %, much higher than the average global growth rate during the same period. Only in 2008 did China post a single-digit growth rate, due to the global financial crisis. The contribution of the automobile industry to the gross domestic product (GDP) in China has increased from 1.50 % in 2005 to 2.61 % in 2010.[2] In 2009, China surpassed Japan to become the top automobile producing country and also topped the United States to become the largest automobile market in the world. Nevertheless, the car penetration rate

[1] "The world's automotive industry," OICA, http://oica.net/wp-content/uploads/2007/06/oica-depliant-final.pdf, 2006.

[2] "The Overall Development Situation and Trend of the Automotive Industry in China," Yang Dong, Shanghai Auto, Vol. 6, 2011.

(CPR, measured as cars per thousand people) is still quite low compared with that of the developed countries, indicating huge potential in this market.

China's automobile manufacturers were originally state-owned, either by the central government, such as FAW and Second Automobile Works (Dongfeng Motors), or by local governments, such as the Beijing Automotive Industry Corporation (BAIC), Chang'an Auto, Guangzhou Auto, and Fujian Auto. With the beginning of the economic reform in 1978, the government started to allow the operation of private automakers and joint ventures with foreign automakers. The first joint venture was Shanghai Volkswagen Automotive Co., Ltd. (SVW), established between Volkswagen (VW) and the Shanghai Automotive Industry Corporation (now SAIC Motor) in 1984. Although there were some conflicts in the process of collaboration, this form of joint venture was widely considered a "win-win" for both sides: foreign automakers could gain entry into the promising Chinese market, lower their production costs through localization of production, and enjoy some policy benefits, and local manufacturers could gain access to the advanced technology, management, and marketing expertise of the foreign partner and capitalize on established foreign brands. In 2011, the top 10 manufacturers sold 6,472,200 cars, accounting for 64 % of total car sales. Eight of them were joint ventures.

As one of the world's leading automobile manufacturers, Toyota Motor Corporation began exporting cars (starting with the Crown sedan) to China in early 1964. However, the localization of production got off to a late start compared with its rivals. Its first joint venture, Tianjin Toyota Motor Engine Co., Ltd., started operating in only 1988. In the same year, Sichuan FAW Toyota Motor Co., Ltd. was founded, and produced the first locally produced Toyota vehicle, the Coaster bus, four years later than its main competitor, the VW Group. Since then, Toyota has hastened its expansion. For each model, Toyota's usual strategy is to import first and then decide whether to localize production depending on the model's popularity. In 2003, the Sichuan plant began to produce the Land Cruiser Prado. In 2004, Toyota established a joint venture, GAC Toyota Motor Co., Ltd., with the Guangzhou Automobile (GAC) Group to produce the Camry in Guangzhou, the capital of Guangdong province and the largest regional automobile market in China. In 2009, Toyota decided to localize the production of its popular SUV model, the Highlander, in GAC-Toyota. By 2011, Toyota had localized the production of 19 sub-brands in China. Its market share in China was 5 %, compared with 18 % in the United States and more than 40 % in Japan. Despite the low market share, the Toyota Camry, RAV4, and Highlander are among China's top sellers in the higher price ranges.

2.2.2 *Toyota China's Production Planning*

To make the best use of Toyota's highly efficient production system, the real challenge lies in accurately forecasting demand and planning production accordingly. This is extremely difficult in a rapidly growing market such as China, particularly for newly introduced models with no historical sales data. Toyota China holds a sales convention at the end of each fiscal year, that gathers major sales representatives from across China. One important mission of this convention is to collect the dealers' replenishment plans for the coming year, which should include both the total quantity and detailed numbers of specific models and configurations for each month. Toyota then determines the yearly quota for each dealer based on the numbers submitted and the consolidated sales in the past year. The general production plan for the coming year is arranged correspondingly. During implementation, Toyota China can make adjustments to the general plan according to the realized sales.

Toyota China has adopted a multi-level management structure. Regional sales managers are responsible for all of the sub-regions within their regions. They determine an overall quota and then allocate the quota among the sub-region dealers. When the demand is realized, they can also arrange transshipments between sub-regions if necessary. This centralized management helps to achieve coordination between dealers. Dealers also enjoy a certain amount of flexibility in demand realization. They must inform Toyota China of their replenishment quantities 3 months in advance, and can then adjust the quantity by up to 10 % until 2 months before delivery, and specify colors up to one month before delivery.

However, although adjustable production and replenishment can help to alleviate the risks of overstock and shortage, the effect is still limited. For instance, the Highlander, a popular SUV model, began local production in China in early 2009. During our field study at GAC-Toyota, we found that in the first half of 2009, the number of Highlanders ordered by dealerships was 60 % higher and the actual realized demand was 90 % higher than Toyota China's projected demand. As some core parts were imported and the capacity was constrained, Toyota China was unable to increase its production in time to fully satisfy the surging demand. Toyota China responded to the supply shortage by expanding production of the Highlander in September 2009, although there was already a long lag. The inevitable result was a prolonged delay in the delivery time to customers: up to 3–6 months compared with the usual one month or less.

2.2.3 Demand Management and Selling Strategies at Dealerships

Like other foreign automakers, the majority of Toyota's marketing, distribution and sales operations in China are conducted by its joint venture dealerships. The dealerships are called 4S stores (sales, spare parts, service, and surveys) and are designed to provide integrated services to customers. According to Toyota, by the end of 2009 it had around 650 4S stores in China, forming the distribution network for its vehicles. 4S stores currently generate most of their profits from new-vehicle sales.

We illustrate the normal operations of a 4S store in Fig. 2.1. The evolutionary Customer Relationship Building (e-CRB) system is an in-house customer relationship management system used by Toyota China's dealerships. Its two key components

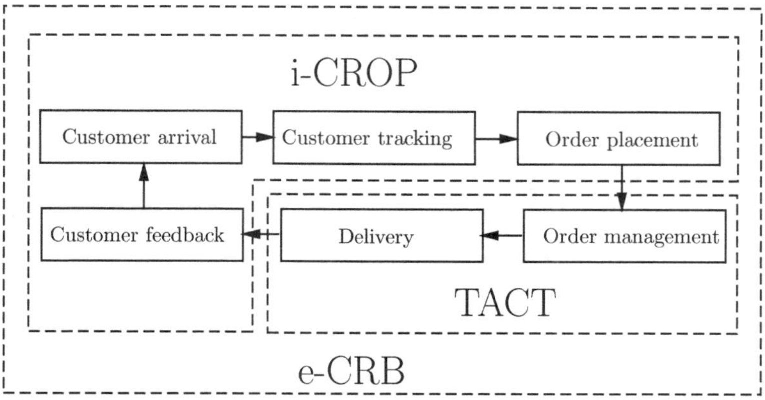

Fig. 2.1 Business process at a 4S store

Table 2.1 Customer categories

Class	Predicted purchasing time
A	1 week
B	Between 1 week and 1 month
C	Between 1 month and 3 months
D	>3 months

Source: Field study at GAC-Toyota

are the intelligent Customer Relationship Optimization Program (i-CROP) and the Total Arranging and Cultivating (TACT) system. The i-CROP is responsible for the management of customer information, and the TACT system is the interface between dealerships and Toyota China. The customer tracking block classifies customers into four classes (A, B, C, and D) according to their intended purchasing times. This classification is usually based on a subjective estimation made by 4S store clerks. A detailed description of the classification system is provided in Table 2.1. Once orders are placed, dealerships use the TACT system to track them until final delivery. According to the 2-month data we collected from a typical 4S store, about 20 % of the in-store customers who purchased the Highlander belonged to classes A and B, and the rest were split evenly between classes C and D.

The interests of Toyota China and its 4S stores are not always perfectly aligned. For example, Toyota China would like to maintain central control over decisions such as pricing, bundling sales, and cross-regional transshipments to avoid image damage and malicious competition, and 4S stores would like to have more flexibility in these decisions. This interest misalignment is particularly prominent when a supply-demand imbalance emerges. As Toyota China cannot fully address the supply shortage by increasing its production, the dealerships must make the best use of their limited inventories to retain customers. In the case of a supply-demand imbalance, dealerships commonly switch to a market segmentation strategy characterized by price and delivery-time differentiation. Two purchasing options are offered to customers: spot and consignment. By choosing the spot option, customers get their desired cars immediately from the on-hand inventory at the expense of paying a higher price, buying a bundled insurance product, or upgrading the configuration. By choosing the consignment option, customers pay the manufacturer's suggested retail price (MSRP) and are put on a waiting list, with delivery usually taking 2–3 months. However, because the "mark-up" associated with the spot option is often determined by dealerships themselves based on the prevailing supply-demand condition, it varies from time to time and from place to place. This has caused a lot of controversy among customers, with some of them even blaming Toyota China for intentionally creating a shortage to rip off customers. Setting aside the controversy, the dealerships that carry out such selling strategies can ameliorate the supply-demand imbalance to some extent; however, they must also deal with extra complexity because they must decide not only whether to accept an order, but also when to fulfill the order.

2.3 Other Examples

Cases similar to the preceding dealership example can be found elsewhere, notably the selling of the iPhone in China.[3] Similar to popular car models, the initial demand for a new iPhone model upon its release can far exceed its supply. As a result, sales must be rationed, and a large proportion of customers are left unfulfilled. However, unlike the dealership example, because Apple itself does not adjust the price or provide customers with multiple delivery options, the market segmentation is carried out by scalpers. Scalpers are usually better at getting newly released iPhones than normal customers. They then immediately put the iPhones they purchased up for sale at a higher price. Hence, a normal customer has two options: he/she can either buy a new iPhone from scalpers right away at a higher price or wait until the initial frenzy fades away, which may take months.

Flexible delivery times may also appear in other forms. For example, it is now a common practice for online retailers such as Amazon.com to offer multiple shipping options to customers. In this case, the scarce resource is the shipping capacity rather than the physical inventory. Shipping capacity is perishable and cannot be carried over. Instead of serving all orders on a priority rules (FCFS) basis, online retailers provide expedited shipping to some priority orders and standard shipping to others. To enjoy the expedited shipping service, however, customers must pay a premium. Compared with the traditional shipping approach, which treats all orders equally, the main merit of multiple shipping options comes from a refined market segmentation based on customers' heterogeneous delivery-time requirements.

In addition to the preceding manufacturing and retailing examples, the application of flexible delivery times can be found in the service industry. The delivery time now corresponds to the delay between when customers arrive and when they are

[3]"China iPhone Craze Breeds Scalpers," Loretta Chao, Wall Street Journal, Sep 30, 2010.

served. A common example is the priority queue system, which is visible at banks, boarding gates, highways, etc.[4] Although a faster service may not be the only reason why customers are willing to pay a premium, it is an important one. Similar to the preceding shipping example, service providers give higher priority to customers who pay to be served ahead of others. Although priority customers can enjoy faster service, normal customers who do not pay are worse off (e.g., they must wait longer). Service providers must decide how to reserve their capacities for different classes of customers.

2.4 Concluding Remarks

It is now well acknowledged that customers are both price and delivery-time sensitive. Although manufacturers and retailers are striving to fulfill customers' orders as soon as possible, there are times when the demand can far exceed the supply and a delay is inevitable. Flexible delivery-time options are an effective means to cope with such situations. As seen in preceding example, because Toyota China is unable to respond to the surging demand in time, dealerships resort to multiple delivery-time options themselves by charging a premium for spot delivery. Even when supply-demand imbalance is not a major concern, firms can make use of flexible delivery times to improve their revenue. As seen in the examples in the retailing and service industries, firms offer multiple delivery-time options to customers or assign higher priority to those customers who are willing to pay a premium to get the service faster.

In all of the examples discussed in this chapter, the firms' optimal strategies are not fixed. That is, both the optimal selling strategy in the dealership example and the optimal capacity allocation strategy in the service-industry examples are contingent on the inventory/capacity status and on realized and future demands. Hence, firms need to adjust their execution of the strategy dynamically. In the following chapters, our theoretical discussions provide an analytical framework for analyzing the preceding examples in terms of how firms should derive their optimal policies dynamically.

The example in Sect. 2.2 is based on Liang et al. (2013).

References

X. Liang, L. Ma, H. Yan, Toyota china: matching supply with demand, in *The Supply Chain Management Casebook: Comprehensive Coverage and Best Practices in SCM*, ed. by C. Munson (Financial Times Press, 2013), pp. 68–79

S. Shugan, Service marketing and management: capacity as a strategic marketing variable, in *The Handbook of Marketing*, ed. by B. Weitz, R. Wensley (Sage Publications, London, 2002), pp. 484–512

[4] "Priority queues: Paying to get to the front of the line," Benjamin Walker, BBC, Oct 10, 2010.

Chapter 3
Inventory Models with Two Delivery-Time Options

Abstract In this chapter, we consider a multi-period inventory model in which a seller provides alternative delivery lead-time choices to customers: a short lead time and a long lead time. In addition to the cyclic inventory replenishment decision, the seller makes the inventory commitment decision, i.e., whether to use a unit of on-hand inventory to satisfy a long lead-time customer immediately. We first characterize the optimal inventory commitment and replenishment policies in a non-capacitated setting. The optimal inventory commitment policy is subject to a commitment level, and the optimal replenishment policy is of a base-stock type. We then consider an extended model with a supply-capacity constraint, in which the commitment policy further includes rejections of short and long lead-time orders. We show that the optimal inventory commitment policy for short lead-time customers is characterized by a switching commitment level and that the optimal inventory commitment policy for long lead-time customers is characterized by three switching commitment levels. We then compare the performance of the optimal dynamic commitment policy with that of a static rationing policy through numerical studies.

3.1 Introduction

In this chapter, we consider a multi-period inventory model in which a seller provides alternative lead-time choices to customers: a short lead time and a long lead time. The model addresses the real-world problem that firms face when serving customers who can be either patient (slow) or impatient (fast). For example, a supplier for clothing retailers may serve both fashion retailers such as H&M and Zara and traditional retailers such as the Gap. Fashion retailers strive to cut their lead times to realize their quick response programs and achieve a shorter operational cycle. Traditional retailers have fewer incentives and encounter more difficulties in adopting quick response programs; thus, they operate with a longer operational cycle (Caro and Martínez-de Albéniz 2010). Hence, the supplier deals with slow and fast orders together. Orders taken by suppliers from slow customers can be included in the next production cycle, and orders from fast customers must be satisfied from the on-hand inventory.

X. Liang et al., *Inventory Management with Alternative Delivery Times*,
SpringerBriefs in Operations Management, DOI 10.1007/978-3-319-48635-2_3

We define a short lead-time customer as one that pays a higher price and requires immediate delivery and a long lead-time customer as one that pays a lower price and accepts a late delivery so that the seller can choose the delivery time at its convenience. Note that the seller has its own lead time on the supply side. We assume that the supply-side lead time is shorter than the lead time of long lead-time customers, but longer than the lead time of short lead-time customers. In addition to the traditional replenishment decision, which is made in cycles, the seller can choose to use the on-hand inventory to satisfy long lead-time customers, which we call an inventory commitment decision. The inventory commitment decision is made on the spot upon the arrival of an order.

We first characterize the optimal inventory commitment and replenishment policies in a non-capacitated setting. We show that the optimal inventory commitment policy is determined by a switching commitment level and that the optimal replenishment policy is of a base-stock type. We then extend our analysis to investigate the case in which there exists a supply capacity constraint. In the capacitated model, we show that the optimal inventory commitment for short lead-time customers is characterized by a threshold level, which determines whether the seller should accept or reject short lead-time customers. We also show that the optimal inventory commitment for long lead-time customers is characterized by three commitment levels. These levels determine whether the seller should accept a long lead-time customer and, if the customer is accepted, whether it should use on-hand inventory to satisfy that customer on the spot. To further illustrate the advantage of the optimal inventory commitment policy, we compare its performance with that of a static inventory rationing policy.

In the remainder of this chapter, Sect. 3.2 presents the notation and formulation of the problem. In Sect. 3.3, we derive the optimal inventory commitment and replenishment policies, first in a non-capacitated setting and then in a capacitated setting. Section 3.4 illustrates the performance of the dynamic commitment policy using numerical examples. Section 3.5 concludes the chapter.

3.2 Notation and Model Formulation

Consider a single-product, finite-time-horizon, periodic-review inventory system that provides two delivery options to its customers: delivery immediately or delivery in the next cycle. At the beginning of each reviewing cycle, the seller replenishes its inventory. Unused inventory is carried over to the next cycle. There are two classes of customers: short and long lead-time customers. When a customer arrives, the seller commits its inventory based on the preference of the customer. If the customer is a short lead-time customer, the seller must satisfy the order with its on-hand inventory. If the customer is a long lead-time customer, the seller can either satisfy the customer now or delay the delivery until the next cycle. Short lead-time customers pay a higher price for the immediate delivery. For now, we assume that the prices of the two delivery options are given. The pricing decision is incorporated and discussed in Chap. 5.

For analytical convenience, following the technique used by Bitran and Mondschein (1997), we divide an inventory cycle into small periods and assume that each period is short enough that at most one customer arrives. Customer arrival is quantified by probabilities. If a short lead-time customer arrives, he/she places an order and expects the order to be delivered immediately. Unfilled demand from short lead-time customers is lost if not satisfied immediately. We assume that there is no cost incurred by the short-demand loss due to stock-out. If a long lead-time customer arrives, he/she places an order and expects the order to be delivered before the next cycle. We start with a non-capacitated model. Hence, all short lead-time customers are satisfied as long as there is still on-hand inventory, and all long lead-time customers can be satisfied at the beginning of the next cycle when the seller replenishes its inventory. At the end of each cycle, the holding cost is calculated, and in the last cycle, the leftover inventory is salvaged.

We introduce the following notation:

$\{1, 2, \ldots, T\}$	the T periods in a cycle;
$n(t)$	the on-hand inventory in period t;
$m(t)$	the outstanding long orders, i.e., the amount of the product promised for delivery by the next cycle in period t;
p_s	the price of the immediate-delivery option;
p_f	the price of the late-delivery option;
π_s	the probability of a short lead-time customer in a period;
π_f	the probability of a long lead-time customer in a period;
π_0	the probability of no customer in a period;
c	the unit purchasing cost;
h	the unit holding cost; and
s	the unit salvage value.

We assume that the following relations hold:

$$p_s > p_f > c > s, \text{ and } \pi_s + \pi_f + \pi_0 = 1.$$

Note that the seller can choose to fulfill long lead-time customers with on-hand inventory. This inventory commitment decision is represented by $u(t)$.

$$u(t) = \begin{cases} 1, & \text{satisfy the long lead-time order with on-hand inventory;} \\ 0, & \text{satisfy the long lead-time order in the next cycle.} \end{cases}$$

To maximize the profit in this inventory system with alternative lead-time options, it is necessary to first derive the optimal dynamic inventory commitment policy $u(t)$, which may be contingent on t, $n(t)$, and $m(t)$. After obtaining the optimal inventory commitment policy, we can continue to address the issue of optimal inventory stocking, i.e., to characterize the optimal inventory replenishment policy.

3.3 Optimal Inventory Commitment and Replenishment Policies

In this section, we first develop the seller's optimal inventory commitment policy within a cycle and then characterize the optimal inventory replenishment policy at the beginning of each cycle. Finally, we revisit these optimal policies in the presence of a supply capacity constraint.

3.3.1 *Optimal Inventory Commitment Policy*

As mentioned earlier, because $p_s > p_f$ and there is no capacity constraint, the seller should always satisfy short lead-time orders as long as the on-hand inventory is non-zero. However, it is critical to time the fulfillment of orders from long lead-time customers. Let $V(t, n, m)$ be the value function in period t with the on-hand inventory level $n(t)$ and the promised-for-delivery level $m(t)$. When $n(t) = 0$, neither long lead-time nor short lead-time customers can be satisfied in the current cycle. As a result, orders from short lead-time customers are lost, and orders from long lead-time customers are scheduled for delivery in the next cycle. The seller obtains a revenue of p_f from each long lead-time customer. Hence, for $1 \leq t < T$,

$$V(t, 0, m) = (\pi_0 + \pi_s)V(t + 1, 0, m) + \pi_f(p_f + V(t + 1, 0, m + 1)),$$

and

$$V(T, 0, m) = -cm.$$

Based on the preceding two equations, a little calculation shows that

$$V(t, 0, m) = (T - t)\pi_f(p_f - c) - cm.$$

When $n(t) \geq 1$, the dynamic programming equation for $1 \leq t < T$ is

$$\begin{aligned} V(t, n, m) =& \pi_0 V(t + 1, n, m) + \pi_f \max_{u \in \{0,1\}} \{V(t + 1, n - u, m + 1 - u)\} \\ &+ \pi_s V(t + 1, n - 1, m) + \pi_f p_f + \pi_s p_s, \end{aligned} \tag{3.1}$$

and the terminal value function is

$$V(T, n, m) = -c(m-n)^{+} - hn + s(n-m)^{+}. \tag{3.2}$$

Eq. (3.1) indicates that to obtain the optimal inventory commitment policy, it is necessary to compare the value functions of two possible actions: using a unit of on-hand inventory to satisfy the long lead-time customer immediately, or promising to deliver in the next cycle. The following lemma provides the monotonicity property of the difference between these two actions. All proofs are relegated to the appendix.

Lemma 3.1 (*a*) *For a given* m, $V(t, n-1, m) - V(t, n, m+1)$ *is non-decreasing in* n*; that is,* $V(t, n-1, m) - V(t, n, m+1) \leq V(t, n, m) - V(t, n+1, m+1)$.

(*b*) *For a given* n, $V(t, n-1, m) - V(t, n, m+1)$ *is independent of* m*; that is,* $V(t, n-1, m) - V(t, n, m+1) = V(t, n-1, m+1) - V(t, n, m+2)$.

Part (*a*) of Lemma 3.1 indicates that the greater the on-hand inventory is, the better it is for the seller to use a unit of on-hand inventory to satisfy the long lead-time customer. Moreover, part (*b*) reveals that this decision is independent of the promised-for-delivery inventory level m. The commitment policy depends on the trade-off between the inventory holding cost for one unit of product and the possible profit from selling it to a short lead-time customer. As the revenue from a long lead-time customer does not change even if the seller delivers the product immediately, the promised-for-delivery level does not play into the tradeoff.

We now present a theorem that characterizes the optimal dynamic inventory commitment policy. Its proof is similar to that of Lemma 3.1. We can obtain the desired results by checking different cases of the term $\max_{u=\{0,1\}}\{V(t, n-u, m+1-u)\}$ in (3.1).

Theorem 3.1 (*a*) *The optimal inventory commitment policy is characterized by a switching commitment level* $C_L(t)$*. The commitment level* $C_L(t)$ *equals* 0 *if* $V(t+1, 0, m) - V(t+1, 1, m+1) \geq 0$*; otherwise, it is defined as the maximum value of* n *such that* $V(t+1, n-1, m) - V(t+1, n, m+1) \leq 0$*. Moreover, the commitment level is independent of the promised-for-delivery level* m *and has a continuity property in relation to the period index* t*, i.e.,* $C_L(t) \leq C_L(t-1) \leq C_L(t) + 1$.

(*b*) *In each period, the optimal control is: to use a unit of the on-hand inventory to satisfy orders from long lead-time customers immediately if* $n(t) > C_L(t)$*; otherwise, to promise to deliver orders from long lead-time customers in the next cycle. That is,*

$$u^*(t) = \begin{cases} 1, & \text{if } n(t) > C_L(t); \\ 0, & \text{otherwise.} \end{cases}$$

Theorem 3.1 characterizes the optimal inventory commitment policy for orders from long lead-time customers. The commitment policy allows the seller to balance the trade-off between using on-hand inventory and using replenished inventory in the next cycle. The former reduces the inventory-holding cost, and the latter reserves on-hand inventory for future short lead-time customers in the rest of the cycle.

3.3.2 Optimal Inventory Replenishment Policy

In our preceding analysis, we obtain the optimal inventory commitment policy for the seller in a selling cycle. We now extend our study to investigate the existence and form of the optimal inventory replenishment policy. We start with a single-cycle problem. We first prove the following lemma.

Lemma 3.2 *The value function $V(t, n, m)$ satisfies the following properties:*

P1 $V(t, n, m)$ *is concave in* n *for a given* m*;*
P2 $V(t, n, m)$ *is concave in* m *for a given* n*; and*
P3 $V(t, n, m)$ *is supermodular in* (n, m).

The supermodularity of $V(t, n, m)$ in relation to (n, m) implies that the on-hand inventory and outstanding long orders are *economic complements*.

Suppose that at the beginning of the cycle the seller has an initial inventory $n(0)$. It then makes a replenishment to bring up the inventory to a desired level $n(1)$, which can be determined as follows:

$$\begin{aligned} &\max_{n(1)\geq n(0)} \{V(1, n(1), 0) - c[n(1) - n(0)]\} \\ &= cn(0) + \max_{n(1)\geq n(0)} \{V(1, n(1), 0) - cn(1)\}. \end{aligned}$$

According to Lemma 3.2, $V(1, n(1), 0)$ satisfies the property of P1. We thus can characterize the optimal inventory replenishment policy in the following theorem.

Theorem 3.2 *The optimal inventory replenishment policy is of a base-stock type. The optimal order-up-to level, denoted by y^*, is the maximizer of $V(1, n(1), 0) - c[n(1) - n(0)]$ and independent of the initial inventory $n(0)$. The optimal ordering quantity, denoted by q^*, is*

$$q^* = \begin{cases} y^* - n(0), & \text{if } n(0) \leq y^*, \\ 0, & \text{if } n(0) > y^*. \end{cases}$$

3.3.3 Optimal Policies with a Supply-Capacity Constraint

In this subsection, we consider the effect of a supply-capacity constraint on the optimal policies. The capacity constraint can be caused by the replenishment lead time. We assume that the replenishment lead time is one cycle. That is, a replenishment order placed in a cycle becomes available in the next cycle. Due to the capacity constraint, orders from long lead-time customers are no longer freely backlogged and satisfied in the next cycle, and orders from short lead-time customers may also be rejected. Hence, the model with a capacity constraint includes the following features: (1) unsatisfied long lead-time orders are subject to a penalty cost, and (2) the inventory commitment policy also considers possibilities of rejecting short and long lead-time orders.

Denote the replenishment quantity and penalty cost for each unfulfilled long lead-time order as Q and c_p, respectively. Let $V^Q(t, n, m)$ be the value function in period t $(1 \leq t < T)$ with the on-hand inventory level $n(t)$ and the promised-for-delivery level $m(t)$.

$$\begin{aligned} V^Q(t,n,m) =& \pi_0 V^Q(t+1,n,m) + \pi_s \max\{V^Q(t+1,n-1,m) + p_s, V^Q(t+1,n,m)\} \\ &+ \pi_f \max\{V^Q(t+1,n-1,m) + p_f, V^Q(t+1,n,m+1) + p_f, \\ &V^Q(t+1,n,m)\}, \end{aligned} \tag{3.3}$$

with a terminal value function

$$V^Q(T,n,m) = -hn + s(n+Q-m)^+ - c_p(m-n-Q)^+. \tag{3.4}$$

Eq. (3.3) indicates that when a long lead-time customer arrives, the seller chooses from three possible actions: using a unit of on-hand inventory to satisfy the customer immediately, promising to deliver the order in the next cycle, or rejecting the order. When a short lead-time customer arrives, the seller chooses whether to use a unit of on-hand inventory to satisfy the customer immediately or reject the order. The following lemma proves the monotonicity properties of the value function.

Lemma 3.3 (*a*) *For a given* m, $V^Q(t+1,n-1,m) - V^Q(t+1,n,m+1)$ *is non-decreasing in* n*; and for a given* n, $V^Q(t+1,n-1,m) - V^Q(t+1,n,m+1)$ *is non-decreasing in* m.

(*b*) *For a given* m, $V^Q(t+1,n-1,m) - V^Q(t+1,n,m)$ *is non-decreasing in* n *(concavity in* n*); and for a given* n, $V^Q(t+1,n-1,m) - V^Q(t+1,n,m)$ *is non-increasing in* m *(supermodular).*

(*c*) *For a given* m, $V^Q(t+1,n,m+1) - V^Q(t+1,n,m)$ *is non-decreasing in* n *(supermodular); and for a given* n, $V^Q(t+1,n,m+1) - V^Q(t+1,n,m)$ *is non-increasing in* m *(concavity in* m*).*

Similar to the results we obtain in the basic model, the supermodularity of $V(t, n, m)$ in (n, m) reveals that the on-hand inventory and promised-for-delivery orders are still *economic complements* in the presence of a capacity constraint. To characterize the optimal inventory commitment policy for long lead-time customers, we introduce the following definitions.

Definition 3.1 (*a*) For each m, a commitment level $C_{L1}(t,m)$ equals 0 if $V^Q(t+1,0,m) - V^Q(t+1,1,m+1) \geq 0$; otherwise, the commitment level is defined as the maximum value of n such that $V^Q(t+1,n-1,m) - V^Q(t+1,n,m+1) \leq 0$.

(*b*) For each m, a commitment level $C_{L2}(t,m)$ equals 0 if $V^Q(t+1,0,m) + p_f - V^Q(t+1,1,m) \geq 0$; otherwise, the commitment level is defined as the maximum value of n such that $V^Q(t+1,n-1,m) + p_f - V^Q(t+1,n,m) \leq 0$.

(*c*) For each m, a commitment level $C_{L3}(t,m)$ equals 0 if $V^Q(t+1,0,m+1) + p_f - V^Q(t+1,0,m) \geq 0$; otherwise, the commitment level is defined as the maximum value of n such that $V^Q(t+1,n,m+1) + p_f - V^Q(t+1,n,m) \leq 0$.

(d) For each m, a commitment level $C_{L4}(t, m)$ equals 0 if $V^Q(t+1, 0, m) + p_s - V^Q(t+1, 1, m) \geq 0$; otherwise, the commitment level is defined as the maximum value of n such that $V^Q(t+1, n-1, m) + p_s - V^Q(t+1, n, m) \leq 0$.

Based on the preceding lemma and definitions, we can characterize the optimal inventory commitment policies for both long and short lead-time customers.

Theorem 3.3 (a) *In period t, for long lead-time customers, the optimal inventory commitment policy is characterized by three switching commitment levels, i.e., $C_{L1}(t, m)$, $C_{L2}(t, m)$, and $C_{L3}(t, m)$; and for short lead-time customers, the optimal inventory-commitment policy is characterized by $C_{L4}(t, m)$.*

(b) *The commitment level $C_{L1}(t, m)$ is non-increasing in m, and commitment levels $C_{L2}(t, m)$, $C_{L3}(t, m)$, and $C_{L4}(t, m)$ are non-decreasing in m.*

(c) $\min\{C_{L1}(t, m), C_{L3}(t, m)\} \leq C_{L2}(t, m) \leq \max\{C_{L1}(t, m), C_{L3}(t, m)\}$. *That is, if $C_{L1}(t, m) = C_{L3}(t, m)$, then $C_{L1}(t, m) = C_{L3}(t, m) = C_{L2}(t, m)$.*

(d) *When $C_{L1}(t, m) > C_{L3}(t, m)$, the optimal inventory commitment policy for long lead-time customers is to satisfy orders immediately if $n(t) > C_{L1}(t, m)$, to promise to deliver orders in the next cycle if $C_{L2}(t, m) < n(t) \leq C_{L1}(t, m)$, and to reject orders otherwise.*

(e) *When $C_{L1}(t, m) \leq C_{L3}(t, m)$, the optimal inventory commitment policy for long lead-time customers is to satisfy orders immediately if $n(t) > C_{L2}(t, m)$, and to reject orders otherwise.*

(f) *The optimal inventory commitment policy for short lead-time customers is to satisfy orders immediately if $n(t) > C_{L4}(t, m)$, and to reject orders otherwise.*

Theorem 3.3 summarizes the optimal inventory commitment policies with a supply-capacity constraint. The optimal policies for short and long lead-time customers are illustrated in Figs. 3.1 and 3.2, respectively. For short lead-time customers, if the on-hand inventory level is high and the promised-for-delivery level is low, the seller has a low risk of running out of the capacity and should satisfy the orders of short lead-time customers using on-hand inventory. If, however, the on-hand inventory level is low and the promsied-for-delivery level is high, the seller should reject the orders of short lead-time customers.

For long lead-time customers, if the on-hand inventory and promised-for-delivery levels are both low, i.e., region I in Fig. 3.2, the seller should promise to deliver the long lead-time order in the next cycle. If the on-hand inventory level is low and the promised-for-delivery level is high, i.e., region II in Fig. 3.2, the seller has a higher chance of running out of capacity and should reject the long lead-time order. If the on-hand inventory level is high, i.e., region III in Fig. 3.2, the seller should satisfy the long lead-time order immediately using on-hand inventory.

In addition, from (c) in Theorem 3.3, we know that the three switching commitment curves intersect at one point because at each switching commitment curve there are two actions leading to the same profit and the seller has three possible actions. This observation implies that we can simplify the control policies by approximating the switching curves with straight lines, as shown by the dotted line in Fig. 3.2. To implement the approximation, we need to locate the following three points: the

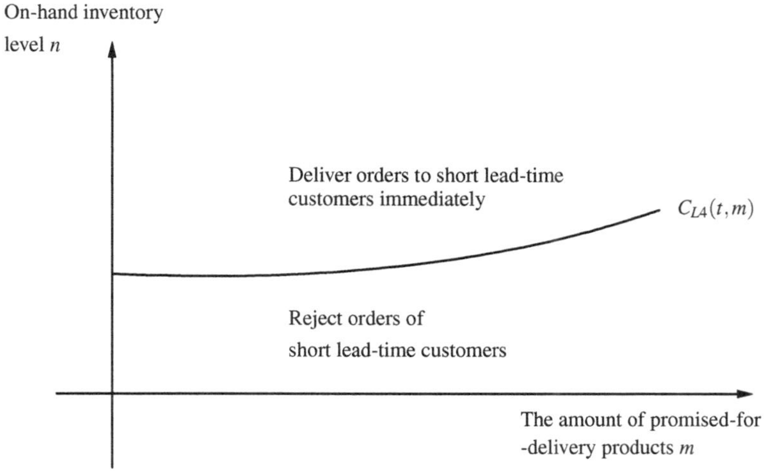

Fig. 3.1 The optimal inventory commitment policy for short lead-time customers with a capacity constraint

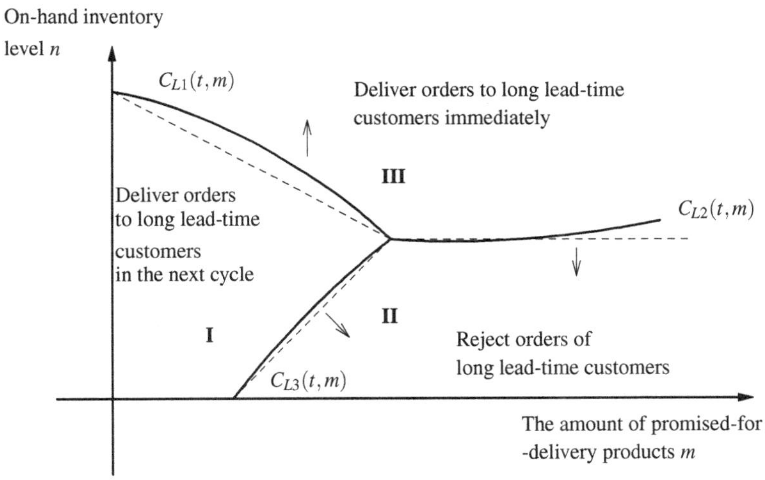

Fig. 3.2 Optimal inventory commitment policy for long lead-time customers with a capacity constraint

intersection of $C_{L1}(t, m)$ with the vertical axis, the intersection of $C_{L3}(t, m)$ with the horizontal axis, and the intersection of the three switching curves. Note that the former two intersections can be calculated by searching the n that makes the two terms $V^Q(t+1, n-1, 0) + p_f$ and $V^Q(t+1, n, 1) + p_f$ equal and the m that makes the two terms $V^Q(t+1, 0, m+1) + p_f$ and $V^Q(t+1, 0, m)$ equal, respectively. The intersection of the three curves can be calculated by searching the (n, m) that makes the three terms $V^Q(t+1, n-1, m) + p_f$, $V^Q(t+1, n, m+1) + p_f$, and $V^Q(t+1, n, m)$ equal.

The following theorem further summarizes the relationship between the optimal inventory commitment policy and the supply capacity Q.

Theorem 3.4 *For a given promised-for-delivery level m,*
(a) $V^Q(t+1, n-1, m) - V^Q(t+1, n, m+1)$ *is non-increasing in* Q.
(b) $V^Q(t+1, n-1, m) - V^Q(t+1, n, m)$ *is non-decreasing in* Q.
(c) $V^Q(t+1, n, m+1) - V^Q(t+1, n, m)$ *is non-decreasing in* Q.
(d) when Q *increases,* $C_{L1}(t, m)$ *does not change or becomes larger (moves upward and rightward in Fig. 3.2, and* $C_{L2}(t, m)$ *and* $C_{L3}(t, m)$ *do not change or become smaller (move downward and rightward in Fig. 3.2.*

When the supply capacity Q increases, the seller is more likely to satisfy orders immediately or promises to deliver orders to the next cycle. As Fig. 3.2 shows, rejection-region II diminishes and promised-for-delivery region I expands. When Q goes to infinity, the capacitated model is reduced to the basic non-capacitated model, and rejection-region II disappears.

3.4 Study of the Inventory Rationing and Inventory Commitment Policies

In Theorem 3.1, we prove that the optimal inventory commitment policy for long lead-time customers can be characterized by a *time-dependent* switching commitment level. In the inventory-rationing literature, it is common to adopt the inventory rationing policy, i.e., to choose a *static* value that defines the maximum number of orders to be reserved for each demand class. Cattani and Souza (2002) compare the performance of such an inventory rationing policy with that of a priority rules policy and show that the former outperforms the latter. In contrast to the static inventory rationing policy, the inventory commitment policy in our model accepts orders based on a dynamic commitment level. In this section, we carry out a study to compare the performance of these two policies.

According to Theorem 3.1, the inventory commitment policy is defined by an optimal commitment level $C_L(t)$ in each period. The commitment level varies in relation to t (see Fig. 3.3 for an illustration). The seller commits one unit of the on-hand inventory to a long lead-time customer as long as the on-hand inventory level is higher than the commitment level. The inventory rationing policy, in contrast, depends on one static inventory rationing level. With the prevailing probabilities of the demand classes listed in Table 3.1, we calculate the profits under the dynamic and static policies. Note that the optimal inventory rationing level is obtained by a complete enumeration. Assume that $p_f = \alpha p_s$ and $\alpha \leq 1$. We use α as the price-decline rate. Let $p_s = 100$, $c = 70$, and $s = 20$. We use the single-cycle model to carry out numerical experiments for different values of h and α.

Let $\alpha = 0.95$. We calculate the optimal inventory commitment levels in Fig. 3.3, which are non-increasing in relation to t. We obtain the optimal profit in relation to the different inventory holding costs in Table 3.2. The inventory rationing policy is

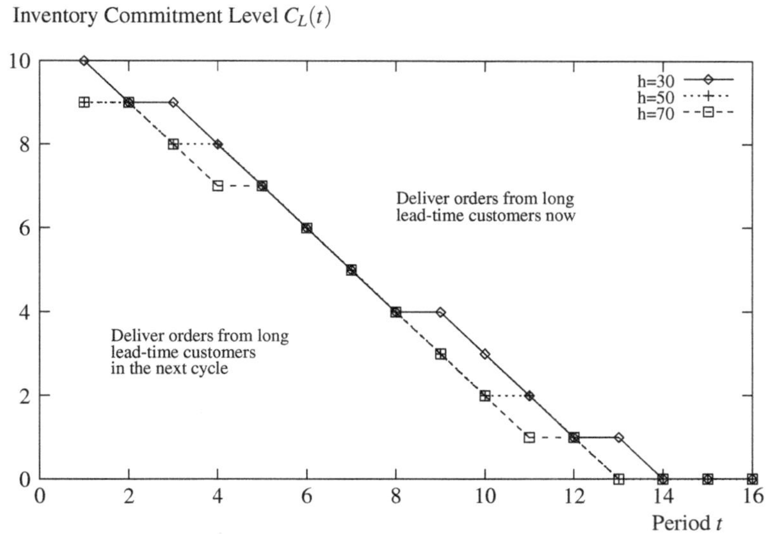

Fig. 3.3 Dynamic inventory commitment levels for different inventory holding costs

Table 3.1 Customer-arrival probabilities in periods

Period	1	2	3	4	5	6	7	8
π_0	0.31	0.18	0.10	0.05	0.03	0.02	0.01	0.02
π_s	0.38	0.43	0.46	0.48	0.49	0.49	0.50	0.49
π_f	0.31	0.39	0.44	0.47	0.48	0.49	0.49	0.49
Period	9	10	11	12	13	14	15	16
π_0	0.03	0.05	0.10	0.18	0.31	0.50	0.70	0.90
π_s	0.49	0.48	0.46	0.43	0.38	0.32	0.22	0.08
π_f	0.48	0.47	0.44	0.39	0.31	0.18	0.08	0.02

implemented by enumerating the rationing level. The improvement ratio is defined as the unified profit difference between the two policies divided by the profit of the inventory rationing policy. Then, we let $h = 50$ and change α to obtain the optimal profits and improvement ratio in Table 3.3. From Tables 3.2 and 3.3, we can see that the optimal profit for the inventory commitment policy is greater than that of the inventory rationing policy, and the improvement ratio is greater than 4 %.

Furthermore, we reduce the horizon length to 8 and obtain the profits for the inventory commitment and inventory rationing policies as 193.91 and 190.24, respectively. The improvement ratio is 1.9 %, which is smaller than the improvement ratio when the horizon length equals 16. This observation is intuitive. As discussed after Theorem 3.1, we indicate that the inventory commitment levels are time-dependent. A smaller t leads to a higher commitment level. Hence, the inventory commitment levels vary in a longer horizon. This demonstrates that the benefit of deploying the inventory commitment policy increases as the length of the cycle increases.

Table 3.2 Profit comparison of the inventory commitment policy and the inventory rationing policy for different h

Inventory holding cost h	Optimal operational profit		
	Rationing policy	Commitment policy	Improvement ratio (%)
70	307.34	319.80	4.1
60	308.09	321.12	4.2
50	308.84	322.45	4.4
40	309.59	324.26	4.7
30	310.34	326.87	5.3

Table 3.3 Profit comparison of the inventory commitment policy and the inventory rationing policy for different α

Price-decline rate α	Optimal operational profit		
	Rationing policy	Commitment policy	Improvement ratio (%)
1	338.65	352.26	4.02
0.98	326.73	340.33	4.16
0.95	308.84	322.45	4.41
0.9	279.03	292.64	4.88
0.83	237.29	250.9	5.74

3.4.1 Robustness of the Inventory Commitment Policy

From Fig. 3.3, the trajectories of the optimal inventory commitment policies seem to suggest that the optimal commitment levels are less sensitive in relation to the holding-cost parameter h. This implies that the inventory commitment policy can be deployed even when the value of the holding cost is uncertain or difficult to obtain. Now, we carry out a numerical experiment to verify this observation. This experiment enables us to quantify the performance changes when an estimated holding-cost parameter is used to derive the inventory commitment policy. Table 3.2 shows that when the inventory-holding cost $h = 60$, the optimal profits for the inventory rationing and inventory commitment policies are 308.09 and 321.12, respectively. We assume that the holding cost for the problem under investigation is $h = 60$. However, the seller uses the dynamic inventory commitment policies derived for $h = 70$ and $h = 50$, and the corresponding profits are 320.65 and 321.10, respectively. Hence, a 1/6 change in the inventory holding cost yields a less than 0.15 % change in the profit. In addition, the dominance over the performance of inventory rationing is maintained. We believe that this observation results for two reasons: the inventory commitment policy dynamically changes the on-hand inventory over the entire cycle, and the ending inventory is consistently lower. As the inventory-holding cost applies only to the ending inventory, it is rational to expect such a performance.

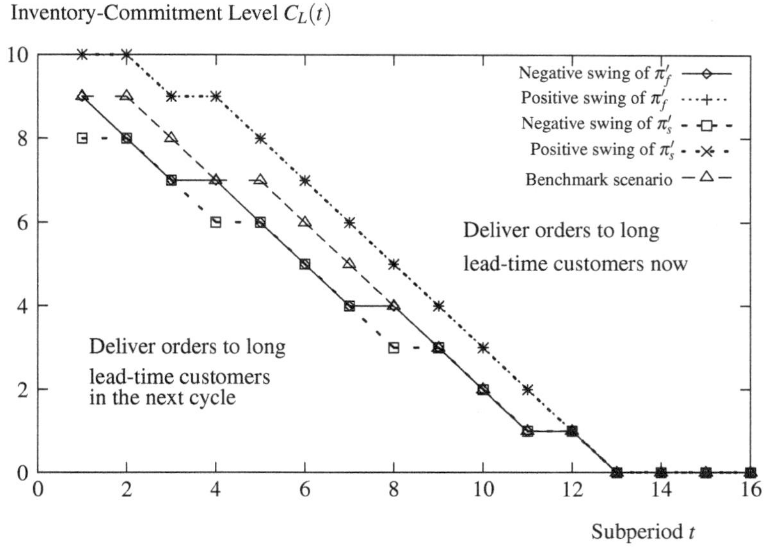

Fig. 3.4 The dynamic inventory commitment levels for different probabilities

Table 3.4 Profits for different policies

Different policies	Fix π_s, change π_f		Fix π_f, change π_s		
	Decrease	Increase	Decrease	Increase	Optimal
Profit	319.22	315.20	318.13	315.20	319.80
Change rate	−0.2 %	−1.4 %	−0.5 %	−1.4 %	–

We also change the demand rates in the periods. We use the demand rates in Table 3.1 as the benchmark probabilities. We first fix π_s and let π_f have a decrease (increase) of $\max\{0.05, \pi'_f\}$ ($\min\{0.05, 1 - \pi'_s\}$). We then fix π_f, and let π_s have a decrease (increase) of $\max\{0.05, \pi'_s\}$ ($\min\{0.05, 1 - \pi'_f\}$). Let $\alpha = 0.95$ and $h = 70$. We compare the inventory commitment levels for the decrease/increase in the long lead-time demand model, the decrease/increase in the short lead-time demand model, and the benchmark case presented in Fig. 3.4. The profits for these inventory-commitment policies are listed in Table 3.4, in which the change rate is equal to the profit difference divided by the optimal profit. The results demonstrate that the degree of profit change is marginal. We believe that the downward slope of the commitment level in relation to time makes the difference here.

3.5 Concluding Remarks

In this chapter, we introduce lead-time choices for customers into the traditional inventory model. Customers are grouped into two classes: one with a short lead-time requirement and the other with a long lead-time requirement. The former asks for a delivery lead time shorter than the lead time of the seller, and the latter asks for a delivery lead time longer than the lead time of the seller. In addition to the inventory replenishment decision, which is made in cycles, the seller makes the inventory commitment decision, i.e., whether to use a unit of on-hand inventory to satisfy a long lead-time customer immediately. We first characterize the optimal inventory commitment and replenishment policies in a non-capacitated setting. We show that the optimal inventory commitment policy is subject to a commitment level and that the optimal replenishment policy is of a base-stock type. We then consider an extended model with a supply-capacity constraint, in which the commitment policy further includes rejections of short and long lead-time orders. We show that in the capacitated model, the optimal inventory commitment policy for short lead-time customers is characterized by a switching commitment level and the optimal inventory commitment policy for long lead-time customers is characterized by three switching commitment levels. We then demonstrate the advantage of the optimal dynamic policy by comparing its performance with that of a static rationing policy through numerical studies.

The inventory commitment policy that allows the seller to satisfy long lead-time customers from on-hand inventory represents the flexibility in the two lead-time options. This flexibility is similar to that seen in the discussion of flexible products (Gallego and Phillips 2004; Chen et al. 2010). In both cases, firms provide multiple options to customers, and different demand classes share some kind of commonality. However, we focus on the role of lead-time differentiation and its implications for inventory decisions in our models. The seller must make two decisions: the inventory replenishment decision, which is made in batches, and the inventory commitment decision, which is made upon the arrival of each customer.

Several limitations of the basic model investigated in this chapter are worth further exploration. For example, we assume that the inventory commitment decision is made on the spot upon the arrival of a customer. However, in some cases, it is also reasonable for the seller to freely choose the time to deliver those backlogged long lead-time orders, as long as it is before the promised delivery time. The prices of the two options are also assumed given in the current model. It is of interest to investigate the effect of the pricing decision on the inventory decisions and system performance. We discuss these extensions in the following chapters.

Appendix

The appendix contains some of the proofs. Some proofs have been omitted to save space and are available upon request from the authors.

Proof of Lemma 3.1

Proof It is straightforward to check that $V(T, n-1, m) - V(T, n, m+1) = -c(m-n+1)^+ - hn + h + s(n-1-m)^+ + c(n+1-n) + hn - s(n-m-1)^+ = h$, which satisfies (a) and (b). Assume that $V(t+1, n-1, m) - V(t+1, n, m+1)$ satisfies (a) and (b). We now prove that $V(t, n-1, m) - V(t, n, m+1)$ also satisfies (a) and (b).

Proof of (a). After checking (3.1), we know that we need only to check the following inequality:

$$\begin{aligned}&\max\{V(t+1, n-2, m), V(t+1, n-1, m+1)\}\\&\quad - \max\{V(t+1, n-1, m+1), V(t+1, n, m+2)\}\\&\leq \max\{V(t+1, n-1, m), V(t+1, n, m+1)\}\\&\quad - \max\{V(t+1, n, m+1), V(t+1, n+1, m+2)\}.\end{aligned} \tag{3.5}$$

We check all possible cases in the remaining proof.

$$\begin{aligned}&V(t+1, n-2, m) - \max\{V(t+1, n-1, m+1), V(t+1, n, m+2)\}\\&\leq V(t+1, n-2, m) - V(t+1, n-1, m+1)\\&\leq V(t+1, n-1, m) - V(t+1, n, m+1)\\&\leq \max\{V(t+1, n-1, m), V(t+1, n, m+1)\} - V(t+1, n, m+1),\end{aligned}$$

in which the second inequality holds because $V(t+1, n-1, m) - V(t+1, n, m+1)$ satisfies (a).

$$\begin{aligned}&V(t+1, n-2, m) - \max\{V(t+1, n-1, m+1), V(t+1, n, m+2)\}\\&\leq V(t+1, n-2, m) - V(t+1, n-1, m+1)\\&\leq V(t+1, n-1, m+1) - V(t+1, n, m+2)\\&\leq V(t+1, n, m+1) - V(t+1, n+1, m+2)\\&\leq \max\{V(t+1, n-1, m), V(t+1, n, m+1)\} - V(t+1, n+1, m+2),\end{aligned}$$

in which the second and third inequalities hold because $V(t+1, n-1, m) - V(t+1, n, m+1)$ satisfies (b) and (a), respectively.

$$
\begin{aligned}
&V(t+1,n-1,m+1) - \max\{V(t+1,n-1,m+1), V(t+1,n,m+2)\} \\
\le\ & 0 \\
\le\ & \max\{V(t+1,n-1,m), V(t+1,n,m+1)\} - V(t+1,n,m+1).
\end{aligned}
$$

$$
\begin{aligned}
&V(t+1,n-1,m+1) - \max\{V(t+1,n-1,m+1), V(t+1,n,m+2)\} \\
\le\ & V(t+1,n-1,m+1) - V(t+1,n,m+2) \\
\le\ & V(t+1,n,m+1) - V(t+1,n+1,m+2) \\
\le\ & \max\{V(t+1,n-1,m), V(t+1,n,m+1)\} - V(t+1,n+1,m+2),
\end{aligned}
$$

in which the second inequality holds because $V(t+1,n-1,m+1)-V(t+1,n,m+2)$ satisfies (a).

From the preceding four inequalities, we know that (3.5) holds.

Proof of (b). After checking (3.1), we know that we need only to check the following equality:

$$
\begin{aligned}
&\max\{V(t+1,n-2,m), V(t+1,n-1,m+1)\} \\
&- \max\{V(t+1,n-1,m+1), V(t+1,n,m+2)\} \\
=&\max\{V(t+1,n-2,m+1), V(t+1,n-1,m+2)\} \\
&- \max\{V(t+1,n-1,m+2), V(t+1,n,m+3)\}.
\end{aligned}
\tag{3.6}
$$

We check all possible cases in the remaining proof.

When $V(t+1,n-2,m) \ge V(t+1,n-1,m+1)$ and $V(t+1,n-1,m+1) \ge V(t+1,n,m+2)$, $V(t+1,n-1,m)-V(t+1,n,m+1)$ satisfies (b), and we know that $V(t+1,n-2,m+1) \ge V(t+1,n-1,m+2)$ and $V(t+1,n-1,m+2) \ge V(t+1,n,m+3)$. Hence, (3.6) is equivalent to

$$
\begin{aligned}
&V(t+1,n-2,m) - V(t+1,n-1,m+1) \\
=\ & V(t+1,n-2,m+1) - V(t+1,n-1,m+2),
\end{aligned}
$$

which is true because $V(t+1,n-2,m) - V(t+1,n-1,m+1)$ satisfies (b).

When $V(t+1,n-2,m) \ge V(t+1,n-1,m+1)$ and $V(t+1,n-1,m+1) \le V(t+1,n,m+2)$, $V(t+1,n-1,m)-V(t+1,n,m+1)$ satisfies (b), and we know that $V(t+1,n-2,m+1) \ge V(t+1,n-1,m+2)$ and $V(t+1,n-1,m+2) \le V(t+1,n,m+3)$. Hence, (3.6) is equivalent to

$$
\begin{aligned}
&V(t+1,n-2,m) - V(t+1,n,m+2) \\
=\ & V(t+1,n-2,m+1) - V(t+1,n,m+3).
\end{aligned}
$$

As $V(t+1,n-2,m) - V(t+1,n-1,m+1)$ satisfies (b), we have $V(t+1,n-2,m)-V(t+1,n,m+2) = V(t+1,n-2,m)-V(t+1,n-1,m+1)+V(t+1,n-1,m+1)-V(t+1,n,m+2) = V(t+1,n-2,m+1)-V(t+1,n-1,m+2)+V(t+1,n-1,m+2)-V(t+1,n,m+3) = V(t+1,n-2,m+1)-V(t+1,n,m+3)$.

When $V(t+1, n-2, m) \leq V(t+1, n-1, m+1)$ and $V(t+1, n-1, m+1) \geq V(t+1, n, m+2)$, $V(t+1, n-1, m) - V(t+1, n, m+1)$ satisfies (b), and we know that $V(t+1, n-2, m+1) \leq V(t+1, n-1, m+2)$ and $V(t+1, n-1, m+2) \geq V(t+1, n, m+3)$. Hence, (3.6) is equivalent to

$$\begin{aligned} &V(t+1, n-1, m+1) - V(t+1, n-1, m+1) \\ &= V(t+1, n-1, m+2) - V(t+1, n-1, m+2), \end{aligned}$$

which is obviously true.

When $V(t+1, n-2, m) \leq V(t+1, n-1, m+1)$ and $V(t+1, n-1, m+1) \leq V(t+1, n, m+2)$, $V(t+1, n-1, m) - V(t+1, n, m+1)$ satisfies (b), and we know that $V(t+1, n-2, m+1) \leq V(t+1, n-1, m+2)$ and $V(t+1, n-1, m+2) \leq V(t+1, n, m+3)$. Hence, (3.6) is equivalent to

$$\begin{aligned} &V(t+1, n-1, m+1) - V(t+1, n, m+2) \\ &= V(t+1, n-1, m+2) - V(t+1, n, m+3), \end{aligned}$$

which is true because $V(t+1, n-1, m+1) - V(t+1, n, m+2)$ satisfies (b).

From the preceding four equalities, we know that (3.6) holds. □

Proof of Theorem 3.1

Proof According to (3.1), it is optimal to deliver now if $V(t+1, n-1, m) - V(t+1, n, m+1) \geq 0$ and to deliver in the next cycle otherwise. Lemma 3.1 provides that in a specific period t, the value of $V(t+1, n-1, m) - V(t+1, n, m+1)$ is non-decreasing in n and independent of m. Hence, if $V(t+1, 0, m) - V(t+1, 1, m+1) \geq 0$, $V(t+1, n-1, m) - V(t+1, n, m+1) \geq 0$ for all $n \geq 1$. In this case, $C_L(t) = 0$, and the optimal policy is to deliver orders to long lead-time customers now for all $n \geq 1$. Otherwise, $C_L(t)$ equals the largest value of n such that $V(t+1, n-1, m) - V(t+1, n, m+1) \leq 0$. $C_L(t)$ is obviously independent of m.

The following is the proof of the non-increasing property of $C_L(t)$ in relation to t. According to the definition of $C_L(t)$, we know that

$$\begin{aligned} &V(t+1, C_L(t)-1, m) \leq V(t+1, C_L(t), m+1), \\ &V(t+1, C_L(t)-2, m) \leq V(t+1, C_L(t)-1, m+1). \end{aligned} \tag{3.7}$$

From (3.1), we know that

$$
\begin{aligned}
&V(t, C_L(t)-1, m) - V(t, C_L(t), m+1) \\
&= \pi_0[V(t+1, C_L(t)-1, m) - V(t+1, C_L(t), m+1)] \\
&+ \pi_f[V(t+1, C_L(t)-1, m+1) - V(t+1, C_L(t), m+2)] \\
&+ \pi_s[V(t+1, C_L(t)-2, m) - V(t+1, C_L(t)-1, m+1)].
\end{aligned}
$$

From (3.7) and Lemma 3.1, we have $V(t+1, C_L(t)-1, m) - V(t+1, C_L(t), m+1) = V(t+1, C_L(t)-1, m+1) - V(t+1, C_L(t), m+2) \leq 0$ and $V(t+1, C_L(t)-2, m) - V(t+1, C_L(t)-1, m+1) \leq 0$. Hence, we know that

$$V(t, C_L(t)-1, m) - V(t, C_L(t), m) \leq 0.$$

As $C_L(t-1)$ is defined as the largest value n such that $V(t, n-1, m) - V(t, n, m) \leq 0$, we know that $C_L(t-1) \geq C_L(t)$.

Also, according to the definition of $C_L(t)$, we have

$$
\begin{aligned}
V(t+1, C_L(t)+1, m) &> V(t+1, C_L(t)+2, m+1), \\
V(t+1, C_L(t), m) &> V(t+1, C_L(t)+1, m+1).
\end{aligned} \tag{3.8}
$$

From (3.7), we know that

$$
\begin{aligned}
&V(t, C_L(t)+1, m) - V(t, C_L(t)+2, m+1) \\
&= \pi_0[V(t+1, C_L(t)+1, m) - V(t+1, C_L(t)+2, m+1)] \\
&+ \pi_f[V(t+1, C_L(t), m) - V(t+1, C_L(t)+1, m+1)] \\
&+ \pi_s[V(t+1, C_L(t), m) - V(t+1, C_L(t)+1, m+1)].
\end{aligned}
$$

From (3.8), we know that

$$V(t, C_L(t)+1, m) - V(t, C_L(t)+2, m) > 0.$$

As $C_L(t-1)$ is defined as the largest value n such that $V(t, n-1, m) - V(t, n, m) \leq 0$, we know that $C_L(t) + 2 > C_L(t-1)$, i.e., $C_L(t) + 1 \geq C_L(t-1)$. □

Proof of Lemma 3.3

Proof We first check the terminal conditions. From (3.4), we have

$$V^Q(T, n-1, m) - V^Q(T, n, m+1) = h,$$

which is non-decreasing in n and m.

$$\begin{aligned}
&V^Q(T,n-1,m)-V^Q(T,n,m)\\
&=h+s(n-1+Q-m)^+-s(n+Q-m)^+-c_p(m-n+1-Q)^++c_p(m-n-Q)^+\\
&=\begin{cases}h-s, & \text{if } n\geq 1-Q+m,\\ h-c_p, & \text{otherwise,}\end{cases}
\end{aligned}$$

which is non-decreasing in n and non-increasing in m.

$$\begin{aligned}
&V^Q(T,n,m+1)-V^Q(T,n,m)\\
&=s(n+Q-m-1)^+-s(n+Q-m)^+-c_p(m+1-n-Q)^++c_p(m-n-Q)^+\\
&=\begin{cases}-s, & \text{if } n\geq 1-Q+m,\\ -c_p, & \text{otherwise,}\end{cases}
\end{aligned}$$

which is non-decreasing in n and non-increasing in m. We assume that $V^Q(t+1,n,m)$ satisfies the properties in the lemma and prove that $V^Q(t,n,m)$ preserves these properties. From (3.3), we know that we need only to focus on the term $\max\{V^Q(t+1,n-1,m), V^Q(t+1,n,m+1), V^Q(t+1,n,m)\}$. In what follows, we prove that $\max\{V^Q(t+1,n-1,m), V^Q(t+1,n,m+1), V^Q(t+1,n,m)\}$ satisfies the properties in this lemma step by step.

We prove that

$$\begin{aligned}
&V^Q(t+1,n-2,m)\\
&\quad-\max\{V^Q(t+1,n-1,m+1), V^Q(t+1,n,m+2), V^Q(t+1,n,m+1)\}\\
&\leq\max\{V^Q(t+1,n-1,m), V^Q(t+1,n,m+1), V^Q(t+1,n,m)\}\\
&\quad-\max\{V^Q(t+1,n,m+1), V^Q(t+1,n+1,m+2), V^Q(t+1,n+1,m+1)\}.
\end{aligned}\tag{3.9}$$

First, we have $V^Q(t+1,n-2,m)-\max\{V^Q(t+1,n-1,m+1), V^Q(t+1,n,m+2), V^Q(t+1,n,m+1)\}\leq V^Q(t+1,n-2,m)-V^Q(t+1,n-1,m+1)\leq V^Q(t+1,n-1,m)-V^Q(t+1,n,m+1)\leq\max\{V^Q(t+1,n-1,m), V^Q(t+1,n,m+1), V^Q(t+1,n,m)\}-V^Q(t+1,n,m+1)$, in which the second inequality holds because $V^Q(t+1,n,m)$ satisfies property (a) of the lemma. Second, we have $V^Q(t+1,n-2,m)-\max\{V^Q(t+1,n-1,m+1), V^Q(t+1,n,m+2), V^Q(t+1,n,m+1)\}\leq V^Q(t+1,n-2,m)-V^Q(t+1,n,m+2)\leq V^Q(t+1,n-1,m)-V^Q(t+1,n+1,m+2)$, in which the last inequality holds because $V^Q(t+1,n,m)$ satisfies property (a) of the lemma. Third, we have $V^Q(t+1,n-2,m)-\max\{V^Q(t+1,n-1,m+1), V^Q(t+1,n,m+2), V^Q(t+1,n,m+1)\}\leq V^Q(t+1,n-2,m)-V^Q(t+1,n,m+1)\leq V^Q(t+1,n-2,m)-V^Q(t+1,n-1,m+1)+V^Q(t+1,n-1,m+1)-V^Q(t+1,n,m+1)\leq V^Q(t+1,n-1,m)-V^Q(t+1,n,m+1)+V^Q(t+1,n,m+1)-V^Q(t+1,n+1,m+1)\leq\max\{V^Q(t+1,n-1,m), V^Q(t+1,n,m+1), V^Q(t+1,n,m)\}-V^Q(t+1,n+1,m+1)$, in which the third inequality holds because $V^Q(t+1,n,m)$ satisfies properties (a) and (b) of the lemma. Thus we prove that (3.9) holds.

We also prove that

$$\begin{aligned} &V^Q(t+1,n-1,m+1) \\ &\quad -\max\{V^Q(t+1,n-1,m+1), V^Q(t+1,n,m+2), V^Q(t+1,n,m+1)\} \\ &\le \max\{V^Q(t+1,n-1,m), V^Q(t+1,n,m+1), V^Q(t+1,n,m)\} \\ &\quad -\max\{V^Q(t+1,n,m+1), V^Q(t+1,n+1,m+2), V^Q(t+1,n+1,m+1)\}. \end{aligned} \tag{3.10}$$

First, we have $V^Q(t+1,n-1,m+1) - \max\{V^Q(t+1,n-1,m+1), V^Q(t+1,n,m+2), V^Q(t+1,n,m+1)\} \le 0 \le \max\{V^Q(t+1,n-1,m), V^Q(t+1,n,m+1), V^Q(t+1,n,m)\} - V^Q(t+1,n,m+1)$. Second, we have $V^Q(t+1,n-1,m+1) - \max\{V^Q(t+1,n-1,m+1), V^Q(t+1,n,m+2), V^Q(t+1,n,m+1)\} \le V^Q(t+1,n-1,m+1) - V^Q(t+1,n,m+2) \le V^Q(t+1,n,m+1) - V^Q(t+1,n+1,m+2) \le \max\{V^Q(t+1,n-1,m), V^Q(t+1,n,m+1), V^Q(t+1,n,m)\} - V^Q(t+1,n+1,m+2)$, in which the second inequality holds because $V^Q(t+1,n,m)$ satisfies property (a) of the lemma. Third, we have $V^Q(t+1,n-1,m+1) - \max\{V^Q(t+1,n-1,m+1), V^Q(t+1,n,m+2), V^Q(t+1,n,m+1)\} \le V^Q(t+1,n-1,m+1) - V^Q(t+1,n,m+1) \le V^Q(t+1,n,m+1) - V^Q(t+1,n+1,m+1) \le \max\{V^Q(t+1,n-1,m), V^Q(t+1,n,m+1), V^Q(t+1,n,m)\} - V^Q(t+1,n+1,m+1)$, in which the second inequality holds because $V^Q(t+1,n,m)$ satisfies property (b) of the lemma. Thus we prove that (3.10) holds.

We also prove that

$$\begin{aligned} &V^Q(t+1,n-1,m) \\ &\quad -\max\{V^Q(t+1,n-1,m+1), V^Q(t+1,n,m+2), V^Q(t+1,n,m+1)\} \\ &\le \max\{V^Q(t+1,n-1,m), V^Q(t+1,n,m+1), V^Q(t+1,n,m)\} \\ &\quad -\max\{V^Q(t+1,n,m+1), V^Q(t+1,n+1,m+2), V^Q(t+1,n+1,m+1)\}. \end{aligned} \tag{3.11}$$

First, we have $V^Q(t+1,n-1,m) - \max\{V^Q(t+1,n-1,m+1), V^Q(t+1,n,m+2), V^Q(t+1,n,m+1)\} \le V^Q(t+1,n-1,m) - V^Q(t+1,n,m+1) \le \max\{V^Q(t+1,n-1,m), V^Q(t+1,n,m+1), V^Q(t+1,n,m)\} - V^Q(t+1,n,m+1)$. Second, we have $V^Q(t+1,n-1,m) - \max\{V^Q(t+1,n-1,m+1), V^Q(t+1,n,m+2), V^Q(t+1,n,m+1)\} \le V^Q(t+1,n-1,m) - V^Q(t+1,n,m+1) \le V^Q(t+1,n,m) - V^Q(t+1,n+1,m+1) \le V^Q(t+1,n,m+1) - V^Q(t+1,n+1,m+2) \max\{V^Q(t+1,n-1,m), V^Q(t+1,n,m+1), V^Q(t+1,n,m)\} - V^Q(t+1,n+1,m+2)$, in which the second and third inequalities hold because $V^Q(t+1,n,m)$ satisfies property (a) of the lemma. Third, we have $V^Q(t+1,n-1,m) - \max\{V^Q(t+1,n-1,m+1), V^Q(t+1,n,m+2), V^Q(t+1,n,m+1)\} \le V^Q(t+1,n-1,m) - V^Q(t+1,n,m+1) \le V^Q(t+1,n,m) - V^Q(t+1,n+1,m+1) \le \max\{V^Q(t+1,n-1,m), V^Q(t+1,n,m+1), V^Q(t+1,n,m)\} - V^Q(t+1,n+1,m+1)$, in which the second inequality holds because $V^Q(t+1,n,m)$ satisfies property (a) of the lemma. Thus, we prove that (3.11) holds.

By combining (3.9), (3.10), and (3.11), we know that

$$\begin{aligned}&\max\{V^Q(t+1,n-2,m),V^Q(t+1,n-1,m+1),V^Q(t+1,n-1,m)\}\\&-\max\{V^Q(t+1,n-1,m+1),V^Q(t+1,n,m+2),V^Q(t+1,n,m+1)\}\\\le&\max\{V^Q(t+1,n-1,m),V^Q(t+1,n,m+1),V^Q(t+1,n,m)\}\\&-\max\{V^Q(t+1,n,m+1),V^Q(t+1,n+1,m+2),V^Q(t+1,n+1,m+1)\}.\end{aligned}\tag{3.12}$$

By applying the same technique, we can prove that

$$\begin{aligned}&\max\{V^Q(t+1,n-2,m),V^Q(t+1,n-1,m+1),V^Q(t+1,n-1,m)\}\\&-\max\{V^Q(t+1,n-1,m+1),V^Q(t+1,n,m+2),V^Q(t+1,n,m+1)\}\\\le&\max\{V^Q(t+1,n-2,m+1),V^Q(t+1,n-1,m+2),V^Q(t+1,n-1,m+1)\}\\&-\max\{V^Q(t+1,n-1,m+2),V^Q(t+1,n,m+3),V^Q(t+1,n,m+2)\}\end{aligned}\tag{3.13}$$

$$\begin{aligned}&\max\{V^Q(t+1,n-2,m),V^Q(t+1,n-1,m+1),V^Q(t+1,n-1,m)\}\\&-\max\{V^Q(t+1,n-1,m),V^Q(t+1,n,m+1),V^Q(t+1,n,m)\}\\\ge&\max\{V^Q(t+1,n-2,m+1),V^Q(t+1,n-1,m+2),V^Q(t+1,n-1,m+1)\}\\&-\max\{V^Q(t+1,n-1,m+1),V^Q(t+1,n,m+2),V^Q(t+1,n,m+1)\}\end{aligned}\tag{3.14}$$

Hence, from (3.12), (3.13), and (3.14), we know that

$$V(t,n-1,m)-V(t,n,m+1)\le V(t,n,m)-V(t,n+1,m+1),\tag{3.15}$$
$$V(t,n-1,m)-V(t,n,m+1)\le V(t,n-1,m+1)-V(t,n,m+2),\tag{3.16}$$
$$V(t,n-1,m+1)-V(t,n,m+1)\le V(t,n-1,m)-V(t,n,m).\tag{3.17}$$

From (3.17), we know that

$$V(t,n-1,m+1)-V(t,n-1,m)\le V(t,n,m+1)-V(t,n,m).\tag{3.18}$$

Furthermore, $V(t,n-1,m)-V(t,n,m)=V(t,n-1,m)-V(t,n,m+1)+V(t,n,m+1)-V(t,n,m)\le V(t,n,m)-V(t,n+1,m+1)+V(t,n+1,m+1)-V(t,n+1,m)$, in which the inequality holds because of (3.15) and (3.18). Hence,

$$V(t,n-1,m)-V(t,n,m)\le V(t,n,m)-V(t,n+1,m).\tag{3.19}$$

In addition, $V(t,n,m+1)-V(t,n,m)=V(t,n,m+1)-V(t,n+1,m+1)+V(t,n+1,m+1)-V(t,n,m)\ge V(t,n,m+2)-V(t,n+1,m+2)+V(t,n+1,m+2)-V(t,n,m+1)$, in which the inequality holds because of (3.19) and (3.16). Hence,

$$V(t,n,m+1)-V(t,n,m)\ge V(t,n,m+2)-V(t,n,m+1).$$

□

Proof of Theorem 3.3

Proof From (a) in Lemma 3.3, we know that $V^Q(t+1, n-1, m) - V^Q(t+1, n, m+1)$ is non-decreasing in both m and n. By definition, $C_{L1}(t, m)$ is, if not 0, the largest value of n such that $V^Q(t+1, n-1, m) - V^Q(t+1, n, m+1) \leq 0$. Hence, we know that $C_{L1}(t, m)$ is non-increasing in m. From (b) and (c) in Lemma 3.3, we can derive that both $C_{L2}(t, m)$ and $C_{L3}(t, m)$ are non-decreasing in m.

From Definition 3.1, we know that for $n(t) > \max\{C_{L1}(t, m), C_{L3}(t, m)\}$, $V^Q(t+1, n-1, m) > \max\{V^Q(t+1, n, m+1), V^Q(t+1, n, m) - p_f\}$, and to deliver orders to long lead-time customers immediately is optimal. In this case, $V^Q(t+1, n-1, m) - V^Q(t+1, n, m) + p_f > 0$, which means $n(t) > C_{L2}(t, m)$. Hence, $C_{L2}(t, m) \leq \max\{C_{L1}(t, m), C_{L3}(t, m)\}$.

For $n(t) \leq \min\{C_{L1}(t, m), C_{L3}(t, m)\}$, $V^Q(t+1, n, m) - p_f \geq \max\{V^Q(t+1, n, m+1), V^Q(t+1, n-1, m)\}$, and it is optimal to reject orders from long lead-time customers. In this case, $V^Q(t+1, n-1, m) - V^Q(t+1, n, m) + p_f \leq 0$, which means $n(t) \leq C_{L2}(t, m)$. Hence, $C_{L2}(t, m) \geq \min\{C_{L1}(t, m), C_{L3}(t, m)\}$.

When $C_{L1}(t, m) > C_{L3}(t, m)$, for $C_{L3}(t, m) < n \leq C_{L1}(t, m)$, $V^Q(t+1, n-1, m) < V^Q(t+1, n, m+1)$ and $V^Q(t+1, n, m+1) > V^Q(t+1, n, m) - p_s$. It is optimal to deliver products to long lead-time customers in the next period.

When $C_{L1}(t, m) < C_{L3}(t, m)$, $C_{L1}(t, m) \leq C_{L2}(t, m) \leq C_{L3}(t, m)$. For $C_{L1}(t, m) < n \leq C_{L2}(t, m)$, $V^Q(t+1, n-1, m) > V^Q(t+1, n, m+1)$ and $V^Q(t+1, n-1, m) < V^Q(t+1, n, m) - p_s$. It is optimal to reject orders from long lead-time customers. For $C_{L2}(t, m) < n \leq C_{L3}(t, m)$, $V^Q(t+1, n-1, m) > V^Q(t+1, n, m) - p_f$ and $V^Q(t+1, n, m+1) < V^Q(t+1, n, m) - p_f$. It is optimal to deliver orders to long lead-time customers immediately. □

Proof of Theorem 3.4

Proof It is easy to prove (a), (b), and (c) by adopting the same technique used in the proof of Lemma 3.3. We prove (d).

When Q increases, from (a), we know that $V^Q(t+1, n-1, m) - V^Q(t+1, n, m+1)$ remains the same or becomes smaller. Recall that $V^Q(t+1, n-1, m) - V^Q(t+1, n, m+1)$ is non-decreasing in n based on Lemma 3.3. Hence, the commitment level $C_{L1}(t, m)$, the biggest value of n such that $V^Q(t+1, n-1, m) - V^Q(t+1, n, m+1) \leq 0$, remains unchanged or increases.

When Q increases, from (b), we know that $V^Q(t+1, n-1, m) - V^Q(t+1, n, m+1)$ remains the same or becomes greater. Recall that $V^Q(t+1, n-1, m) - V^Q(t+1, n, m+1)$ is non-decreasing in n based on Lemma 3.3. Hence, the commitment level $C_{L2}(t, m)$, the biggest value of n such that $V^Q(t+1, n-1, m) + p_f - V^Q(t+1, n, m) \leq 0$, remains unchanged or decreases.

When Q increases, from (c), we know that $V^Q(t+1, n-1, m) - V^Q(t+1, n, m+1)$ remains the same or becomes greater. Recall that $V^Q(t+1, n-1, m) - V^Q(t+1, n, m+1)$ is non-decreasing in n based on Lemma 3.3. Hence, the commitment level $C_{L3}(t, m)$, the biggest value of n such that $V^Q(t+1, n, m+1) + p_f - V^Q(t+1, n, m) \leq 0$, remains unchanged or decreases. □

References

G.R. Bitran, S.V. Mondschein, Periodic pricing of seasonal products in retailing. Manag. Sci. **43**(1), 64–79 (1997)

F. Caro, V. Martínez-de, Albéniz. The impact of quick response in inventory-based competition. Manuf. Serv. Oper. Manag. **12**(3), 409–429 (2010)

K.D. Cattani, G.C. Souza, Inventory rationing and shipment flexibility alternatives for direct market firms. Prod. Oper. Manag. **11**(4), 441 (2002)

S. Chen, G. Gallego, Z. Li, B. Lin, Optimal seat allocation for two-flight problems with a flexible demand segment. Eur. J. Oper. Res. **201**(3), 897–908 (2010)

G. Gallego, R. Phillips, Revenue management of flexible products. Manuf. Serv. Oper. Manag. **6**(4), 321–337 (2004)

Chapter 4
Inventory Models with Delivery-Time Upgrade

Abstract In this chapter, we extend the unitary inventory commitment policy discussed in Chap. 3 to a more general upgrade policy, which allows the seller to choose the delivery time of all backlogged long lead-time orders. We characterize the optimal inventory upgrade and replenishment policies. In our analysis, we resort to the technique of anti-multimodularity to derive the structural properties of the value function. We further calibrate the model parameters using the empirical data collected from Toyota China dealerships and calculate the optimal policies and several performance metrics accordingly. The numerical results show that the upgrade mechanism can improve both profit and inventory usage.

4.1 Introduction

In Chap. 3, we introduce our formulation of inventory models with alternative delivery lead times. Between inventory replenishments, the seller makes use of inventory commitment control to handle incoming orders from both short and long lead-time customers. It is assumed that the inventory commitment decision is made on the spot. In this chapter, we consider a generalization of this commitment decision. Although the rejection decisions (applicable in the capacitated setting) are still made on the spot, the early delivery of a backlogged long lead-time order, i.e., the delivery-time upgrade, can be carried out at any time before the promised delivery time.

The upgrade mechanism is consistent with the business practice at Toyota China dealerships, which we introduce in detail in Chap. 2. As previously mentioned, the forecast by Toyota China is plagued by the problem of high volatility in the Chinese automobile market. For instance, Highlander, a popular SUV produced by Toyota, began local production in China in early 2009. Compared with Toyota's projected demand in the first half of 2009, the order quantity placed by dealerships was 60 % higher and the actual realized demand was 90 % higher. Toyota China responded to the supply shortage by beginning to expand its production in September 2009, which was already very late. Hence, a specific strategy characterized by differentiated prices and delivery times is typically used at dealerships (known as 4S stores: sales, service, spare parts, and surveys) to handle the supply-demand imbalance. In general,

X. Liang et al., *Inventory Management with Alternative Delivery Times*,
SpringerBriefs in Operations Management, DOI 10.1007/978-3-319-48635-2_4

when demand exceeds supply, dealerships begin to sell cars in two forms: spot and consignment. Customers who choose the spot option get their desired cars from the stock at a higher price, and customers who choose the consignment option pay the list price and join the waiting list. The orders of consignment customers usually take two to three months to clear. To implement this strategy, dealerships must decide whether to fulfill or delay an incoming order, in addition to the backlogged long lead-time orders, with updated information about inventory and demand. The aim of our theoretical analysis of the upgrade mechanism is to develop models, derive management strategies, and reveal insights into this market-segmentation strategy.

The model formulation and analysis follow those of the basic model in Chap. 3, with a few notable differences. First, the upgrade decision generalizes the unitary inventory control in the basic model. It allows the seller to determine the delivery of not only the incoming order in the current period but also previously backlogged long lead-time orders. Second, to facilitate our analysis, we make a variable substitution to replace the outstanding long lead-time orders by the net inventory, which equals the on-hand inventory level minus the number of promised-for-delivery orders. Third, we resort to the technique of anti-multimodularity to derive the structural properties of the value function and obtain the optimal policies in the capacitated setting. In addition to the preceding changes in our theoretical analysis, we calibrate the model parameters using the empirical data collected from Toyota China dealerships and calculate the optimal policies and several performance metrics accordingly. The results show the improvements in both profit and inventory usage with the adoption of the upgrade mechanism.

In the remainder of this chapter, Sect. 4.2 presents the formulation of the problem. In Sect. 4.3, we obtain the optimal inventory allocation, upgrade, and replenishment policies and discuss their properties, first in a non-capacitated setting and then in a capacitated setting. In Sect. 4.4, we conduct numerical studies to calculate the optimal policies and demonstrate the performance of the upgrade mechanism using parameters calibrated from empirical data. Section 4.5 concludes the chapter.

4.2 Model Formulation

The model setup and notation mostly follow those of the basic model in Chap. 3. We still consider a single-product, finite-time-horizon, periodic-review inventory system. Inventory is replenished at the beginning of each review cycle, and unused inventory is carried over to the next cycle. A cycle is further divided into T periods, and a period is short enough such that at most one customer arrives in each period. The seller provides two delivery-time options to its customers: immediate delivery (short) or delivery at the beginning of the next cycle (long). Accordingly, there are two classes of customers: short and long lead-time customers. Their arrival probabilities in a period remain the same as in the basic model.

A short lead-time customer requires an immediate delivery. The order is lost if the seller runs out of on-hand inventory. For a long lead-time customer, the seller can either fulfill his/her order at the next replenishment as promised or fulfill it earlier, including immediate delivery on the spot. In each period, the seller also reviews previously backlogged long lead-time orders and determines whether to delivery them in the current period. We denote the action of fulfilling demand from long lead-time customers with on-hand inventory as an *upgrade*. We assume that long lead-time customers prefer an early delivery and thus always accept the upgrade.

The cost parameters remain the same as in the basic model. However, the holding cost is now calculated at the end of each period rather than at the end of the cycle. In the basic model, the inventory commitment decision can determine only the early delivery of a long lead-time order upon its arrival. The calculation of the holding cost thus can be postponed to the end of the cycle. The upgrade decision, in contrast, can choose a more flexible delivery time. It is therefore necessary to calculate the holding cost in each period. Otherwise, the seller can choose to delay all deliveries until the last period in a cycle.

In period t, the seller chooses an upgrade quantity, denoted by $u(t)$. The upgrade quantity $u(t)$ is constrained by the on-hand inventory level $n(t)$ and outstanding long orders $m(t)$. We denote the feasible range of $u(t)$ for given $n(t)$ and $m(t)$ as $\mathscr{U}(n, m)$. Let $\mathbb{Z}^+$ be the set of non-negative integers. We have

$$\mathscr{U}(n, m) = \{u : 0 \leq u \leq n \wedge m, u \in \mathbb{Z}^+\},$$

where $n \wedge m = \min\{n, m\}$. In each period, there are three possibilities in relation to customer arrival: a short lead-time customer, a long lead-time customer, or no customer. We denote the upgrade quantities in the three scenarios as u_s, u_f, and u_0, respectively. We now charaterize the inventory dynamics. When $n(t) > 0$,

$$\begin{aligned}&(n(t+1), m(t+1))\\ =& \begin{cases} (n(t) - 1 - u_s, m(t) - u_s), & \text{if a short lead-time customer arrives;} \\ (n(t) - u_f, m(t) + 1 - u_f), & \text{if a long lead-time customer arrives;} \\ (n(t) - u_0, m(t) - u_0), & \text{no customer arrives.} \end{cases}\end{aligned}$$

When $n(t) = 0$,

$$(n(t+1), m(t+1)) = \begin{cases} (0, m(t) + 1), & \text{if a long lead-time customer arrives;} \\ (0, m(t)), & \text{if a short lead-time or no customer arrives.} \end{cases}$$

We start with a single-cycle model without a supply-capacity constraint. As there is no capacity constraint, the seller satisfies all short lead-time orders and accepts all long lead-time orders. Let $V(t, n, m)$ be the value function in period t. For $1 \leq t < T$, if $n(t) = 0$, the short lead-time order is lost, and the long lead-time order is backlogged for future delivery. Hence, the dynamic programming recursion can be written as follows:

$$V(t, 0, m) = (\pi_0 + \pi_s)V(t + 1, 0, m) + \pi_f[p_f + V(t + 1, 0, m + 1)].$$

If $n(t) \geq 1$, the seller can choose whether to upgrade outstanding long lead-time orders and how many to upgrade. We have

$$\begin{aligned} V(t,n,m) = \pi_0 & \max_{u_0 \in \mathscr{U}(n,m)} V(t+1, n-u_0, m-u_0) \\ & + \pi_f[\max_{u_f \in \mathscr{U}(n,m+1)} V(t+1, n-u_f, m+1-u_f) + p_f] \\ & + \pi_s[\max_{u_s \in \mathscr{U}(n-1,m)} V(t+1, n-1-u_s, m-u_s) + p_s] - hn, \end{aligned}$$

and

$$V(T,n,m) = -c(m-n)^+ + s(n-m)^+.$$

A close look at the preceding dynamic programming equations reveals that the terminal function and upgrade decision are more involved with the difference between the on-hand inventory level and outstanding long lead-time orders, i.e., $n(t) - m(t)$. Further analysis indicates that it is mathematically convenient to use this difference variable rather than $m(t)$. We call the difference variable *the net inventory* and denote it as $m_n(t)$. The feasible range of $u(t)$ and the dynamic programming equations can be updated as follows:

$$\mathscr{U}(n(t), m_n(t)) = \{u : 0 \leq u \leq n(t) \wedge (n(t) - m_n(t)), u \in \mathbb{Z}^+\},$$

$$\begin{aligned} V(t,n,m_n) = \; & \pi_0 \max_{u_0 \in \mathscr{U}(n,m_n)} V(t+1, n-u_0, m_n) \\ & + \pi_f \max_{u_f \in \mathscr{U}(n,m_n-1)} V(t+1, n-u_f, m_n-1) \\ & + \pi_s \max_{u_s \in \mathscr{U}(n-1,m_n-1)} V(t+1, n-1-u_s, m_n-1) \\ & + \pi_f p_f + \pi_s p_s - hn, \\ V(t,0,m_n) = \; & (\pi_0 + \pi_s) V(t+1, 0, m_n) \\ & + \pi_f[p_f + V(t+1, 0, m_n-1)], \end{aligned}$$

and

$$V(T,n,m_n) = -cm_n^- + sm_n^+.$$

The upgrade decision reduces on-hand inventory and outstanding long lead-time orders at the same time and thus does not change the value of the net inventory. Note that the net inventory can be either positive or negative. A negative net inventory means that the on-hand inventory is insufficient to satisfy all backlogged long lead-time orders.

4.3 Optimal Inventory Upgrade and Replenishment Policies

In this section, we first characterize the seller's optimal inventory upgrade and replenishment policies without a supply-capacity constraint. We then derive the optimal policies in the presence of a supply-capacity constraint.

4.3.1 Optimal Inventory Upgrade Policy

To derive the optimal upgrade policy, we first introduce the following preliminary results. Proofs are relegated to the appendix.

Lemma 4.1 *(a) For a given m_n, $V(t, n-1, m_n) - V(t, n, m_n)$ is non-decreasing in n, i.e., $V(t, n-1, m_n) - V(t, n, m_n) \leq V(t, n, m_n) - V(t, n+1, m_n)$.*
(b) $V(t, n-1, m_n) - V(t, n, m_n)$ is independent of m_n, i.e., $V(t, n-1, m_n) - V(t, n, m_n) = V(t, n-1, m_n+1) - V(t, n, m_n+1)$.

Part (a) of Lemma 4.1 indicates that given a net inventory level, the marginal profit of on-hand inventory, $V(t, n, m_n) - V(t, n-1, m_n)$, is non-increasing as on-hand inventory increases. Part (b) further shows that the marginal profit of on-hand inventory is independent of the net inventory level. With the properties in Lemma 4.1, we characterize the optimal upgrade policy in the following theorem.

Theorem 4.1 *(a) In period t, the optimal inventory upgrade policy is characterized by a threshold level $C_L(t)$. It equals 0 if $V(t, 0, m_n) - V(t, 1, m_n) > 0$ and equals $m_n(t)$ if $V(t, m, m) - V(t, m+1, m) > 0$. Otherwise, $C_L(t)$ is defined as the maximum value of $n(t)$ such that $V(t, n-1, m_n) - V(t, n, m_n) \leq 0$. Moreover, $C_L(t)$ is independent of the net inventory $m_n(t)$ in the last case.*
(b) In period t, for a given $m_n(t)$, if $n(t) > C_L(t)$, upgrade down to $C_L(t)$. If $n(t) < C_L(t)$, make no upgrade.

Lemma 4.1 and Theorem 4.1 respectively resemble Lemma 3.1 and Theorem 3.1 in Chap. 3, which are obtained in the case of unitary inventory commitment control. The inventory upgrade policy now gives greater flexibility to the seller, and the optimal upgrade policy that we describe in Theorem 4.1 can be viewed as a reverse base-stock policy. That is, rather than order up to, the seller chooses to upgrade its on-hand inventory level down to the optimal threshold. We look at the three upgrade quantities: u_0, u_s, and u_f. In period t, the following relations hold: if $u_0 > 0$, $u_0 = u_f = u_s + 1$, and if $u_0 = 0$, $u_0 = u_f = u_s = 0$. These simple relations come directly from the properties in Lemma 4.1.

4.3.2 Optimal Inventory Replenishment Policy

We now investigate the optimal inventory replenishment policy at the beginning of the cycle. In addition to the properties in Lemma 4.1, we can obtain the following structural property.

Lemma 4.2 *$V(t, n, m_n)$ is integer concave in $m_n(t)$.*

At the beginning of the cycle, we assume that the seller starts with an initial inventory $n(0)$. The inventory replenishment decision can be written as follows.

$$\begin{aligned}&\max_{n(1)\geq n(0)}\{V(1, n(1), n(1)) - c[n(1) - n(0)]\}\\ =\ & cn(0) + \max_{n(1)\geq n(0)}\{V(1, n(1), n(1)) - cn(1)\}.\end{aligned}$$

As there is no outstanding long lead-time order at the beginning of the cycle, the net inventory equals the on-hand inventory, i.e., $m_n(1) = n(1)$. $V(1, n(1), n(1))$ is concave in $n(1)$ given the integer concavities and modularity in n and m_n. This makes it possible to characterize the optimal inventory replenishment policy.

Theorem 4.2 *The inventory replenishment policy is of a base-stock type. There exists an optimal order-up-to level $n^*(1)$ such that the optimal ordering quantity is*

$$q^* = \begin{cases} n^*(1) - n(0), & if\, n(0) \leq n^*(1),\\ 0, & if\, n(0) > n^*(1),\end{cases}$$

where $n^*(1)$ is the maximizer of $V(1, n(1), n(1)) - c[n(1) - n(0)]$.

4.3.3 Multi-cycle Problem

Based on the results derived in the single-cycle problem, we are ready to extend our discussion to the general multi-cycle problem. In the multi-cycle problem, the seller replenishes its inventory at the beginning of each cycle. This replenishment serves two purposes: to clear the remaining outstanding long lead-time orders from previous cycle and to hold stock for the demand in the upcoming cycle.

The derivation of the optimal policies still depends on the following three key properties.

Integer concavity in n: $V(t, n, m_n) - V(t, n-1, m_n) \geq V(t, n+1, m_n) - V(t, n, m_n)$;

Integer concavity in m_n: $V(t, n, m_n) - V(t, n, m_n - 1) \geq V(t, n, m_n + 1) - V(t, n, m_n)$;

Modularity in n and m_n: $V(t, n, m_n) - V(t, n-1, m_n) = V(t, n, m_n - 1) - V(t, n-1, m_n - 1)$.

In preceding analysis, we prove that these three properties hold within a single cycle. For the multi-cycle problem, we further prove that the inventory replenishment at the beginning of each cycle preserves these properties.

Lemma 4.3 *The optimal inventory replenishment policy at the beginning of each cycle preserves the following properties: the integer concavities and modularity in* n *and* m_n.

Given that the three properties hold for all $V(t, n, m_n)$, the results derived in the single-cycle problem still apply.

Theorem 4.3 *For the multi-cycle problem, the optimal upgrade policy is of a threshold type, as described in Theorem 4.1, and the optimal inventory replenishment policy at the beginning of each cycle follows a base-stock policy, as described in Theorem 4.2.*

Recall that in Theorem 3.1 in Chap. 3, we prove that the commitment level decreases over time and the decrement is at most one unit per period. We next show that this property remains true in the upgrade model, as stated in the following theorem.

Theorem 4.4 *When* h *is time-invariant, denote the optimal thresholds for two consecutive periods* t *and* $t+1$ *within a cycle as* $C_L(t)$ *and* $C_L(t+1)$*, respectively. We have*

$$0 \leq C_L(t) - C_L(t+1) \leq 1.$$

Note that in the preceding discussion and dynamic programming equations, we simply use h to represent the holding cost. It is obvious that h could be time-dependent, i.e., as a function of time, $h(t)$, and all preceding results still hold. However, to derive the result in Theorem 4.4, the assumption that h is time invariant is necessary. When this assumption is violated, the simple structure of the threshold may not hold. That is, the threshold $C_L(t)$ may increase rather than decrease over time, and the change per period can be greater than one unit.

With the property of $C_L(t)$ established in Theorem 4.4, the dynamic programming equation can be simplified. For a seller, starting with zero outstanding long lead-time order, the optimal action in any period is to choose whether to upgrade one outstanding long lead-time order if a long lead-time customer or no customer arrives in that period. Otherwise, no action needs to be taken.

4.3.4 Optimal Policies with a Supply-Capacity Constraint

In this subsection, we incorporate a supply-capacity constraint into the upgrade model. We assume that there is a capacity constraint for the replenishment at the beginning of each cycle, denoted by Q. The supply capacity is not imposed as a hard constraint. When a replenishment exceeds the capacity, a penalty cost, denoted as c_p, is incurred for each extra unit. As a result, to avoid the penalty, the seller can choose to reject short and long lead-time orders. It is clear that when $c_p = 0$, the problem is reduced to the non-capacitated case.

Recall that we define the upgrade quantities in three possible demand-arrival scenarios as u_s, u_f, and u_0, respectively in the non-capacitated model. Due to the incorporation of rejection decisions, u_s, u_f, and u_0 are now better understood as the upgrade quantities that respectively correspond to three possible system states. The three states are generated by the customer arrival and the seller's subsequent acceptance/rejection decision. We explain them in detail here. Assume that the system state is (n, m_n) at the beginning of period t. If no customer arrives, the seller chooses an upgrade quantity u_0, and the state changes to $(n - u_0, m_n)$. If a short lead-time customer arrives, the seller first decides whether to accept or reject the order. If the order is accepted, the seller chooses an upgrade quantity u_s, and the state changes to $(n - 1 - u_s, m_n - 1)$; if the order is rejected, the seller chooses an upgrade quantity u_0, and the state changes to $(n - u_0, m_n)$. If a long lead-time customer arrives, the seller decides whether to accept or reject the order. If the order is accepted, the seller chooses an upgrade quantity u_f, and the state changes to $(n - u_f, m_n - 1)$; if the order is rejected, the seller chooses an upgrade quantity u_0, and the state changes to $(n - u_0, m_n)$. To further simplify the notation, we use u_s to represent u_f by noting that in the long-lead-time-demand scenario, $u_f = 0$ if the seller chooses not to upgrade and $u_f = u_s + 1$ if the seller upgrades.

We still start with a single-cycle problem. Let $V^Q(t, n, m_n)$ be the value function in period t. The dynamic programming equations can be rewritten as follows:

$$
\begin{aligned}
V^Q(t,n,m_n) =& \pi_0 \max_{u_0 \in \mathcal{U}(n,m_n)} V^Q(t+1, n-u_0, m_n) \\
&+ \pi_s \max_{\substack{u_0 \in \mathcal{U}(n,m_n), \\ u_s \in \mathcal{U}(n-1,m_n-1)}} \left\{ V^Q(t+1, n-1-u_s, m_n-1) + p_s, V^Q(t+1, n-u_0, m_n) \right\} \\
&+ \pi_f \max_{\substack{u_0 \in \mathcal{U}(n,m_n), \\ u_s \in \mathcal{U}(n-1,m_n-1)}} \{ V^Q(t+1, n-1-u_s, m_n-1) + p_f, \\
&\quad V^Q(t+1, n, m_n-1) + p_f, V^Q(t+1, n-u_0, m_n) \} - (b+h)n + bm_n, \\
V^Q(t,0,m_n) =& (\pi_0 + \pi_s) V^Q(t+1, 0, m_n) + \pi_f \max \left\{ p_f + V^Q(t+1, 0, m_n-1), V^Q(t+1, 0, m_n) \right\} \\
&+ bm_n, \\
V^Q(T,n,m_n) =& sm_n^+ - cm_n^- - c_p(-m_n - Q)^+.
\end{aligned}
$$

To understand the structural properties of the value function, we introduce the following definition.

Definition 4.1 (Altman et al. 2003; Hajek 1985) Let $e_i \in \mathbb{N}^k$ $(i = 1, \cdots, k)$ denote the vector with all entries equal to 0 except for a 1 in its ith entry. Define $s_i = e_{i-1} - e_i, i = 2, \cdots, k$. Let $\mathcal{F} = \{-e_1, s_2, s_3, \ldots, s_{k-1}, s_k, e_k\}$. Then, a real-valued function $f : \mathbb{Z}^k \to \mathbb{R}$ is multimodular in relation to $\mathcal{F}$ if for all $x \in \mathbb{Z}^k$, v and w in $\mathcal{F}, v \neq w$, and the following holds: $f(x+v) + f(x+w) \geq f(x) + f(x+v+w)$.

Multimodularity is one way to define joint convexity for multiple discrete variables. Interested readers may refer to Altman et al. (2003) for a thorough treatment of this topic. For our problem, we need the following property of anti-multimodularity.

Definition 4.2 (Narongwanich et al. 2002) A real-valued function $f : \mathbb{Z}^k \to \mathbb{R}$ is said to be anti-multimodular if $-f$ is multimodular.

For brevity, in what follows we refer to anti-multimodularity/anti-multimodular as AMM. We can prove the following lemma.

Lemma 4.4 *$V^Q(t, n, m_n)$ is AMM on $(-n, m_n)$.*

We provide a sketch of the proof here. For an AMM function, we can construct a continuously concave counterpart through linear interpolation (Altman et al. 2003). To prove the AMM is preserved by the dynamic programming recursion, we first transform the value function into its continuous counterpart and prove the preservation of joint concavity of the new continuous function. We then prove that the transformation is reversible. One advantage of such a proof procedure is that it proves all of the discrete properties together rather than one by one by using the discrete-continuous-discrete transformation.

We are now ready to characterize the seller's optimal inventory control policy based on the preceding structural properties. In each period, depending on the customer arrival, the seller makes two sequential decisions: whether to accept or reject the incoming short (long) lead-time order, if any, and then to choose an upgrade quantity for the outstanding long lead-time orders. In period t, the upgrade decision depends on the value of $V^Q(t+1, n, m_n) - V^Q(t+1, n-1, m_n)$. As long as this value is positive, the seller should keep upgrading. Define $C_L(t, m_n)$ as $C_L(t, m_n) = \arg\max_n \{V^Q(t+1, n, m_n) - V^Q(t+1, n-1, m_n) \geq 0\}$. If such an $n(t)$ does not exist, $C_L(t, m_n) = \max\{m_n(t), 0\}$. Following the AMM, we know that $C_L(t, m_n)$ is non-decreasing in $m_n(t)$ and $C_L(t, m_n+1) - C_L(t, m_n) \leq 1$. On the right side of $C_L(t, m_n)$ (i.e., $n(t) > C_L(t, m_n)$), it is optimal for the seller to upgrade down to $C_L(t, m_n)$. On the left side of $C_L(t, m_n)$ (i.e., $n(t) < C_L(t, m_n)$), the seller makes no upgrade.

A similar logic applies to the handling of an incoming short or long lead-time order. For a short lead-time order, the seller's action depends on the value of $V^Q(t, n, m_n) - V^Q(t, n-1, m_n-1) - p_s$. The seller should fulfill the order as long as this value is negative and reject it otherwise. On the left side of $C_L(t, m_n)$, we define $C_S(t, m_n)$ as $C_S(t, m_n) = \arg\max_n \{V^Q(t, n, m_n) - V^Q(t, n-1, m_n-1) \geq p_s\}$. According to the AMM, $C_S(t, m_n)$ is non-increasing in $m_n(t)$. On the right side of $C_L(t, m_n)$, because of the subsequent upgrade decision, $C_S(t, m_n)$ becomes a line parallel to the n-axis, which implies that it does not change with $n(t)$. Its value is defined as $C_S(t) = \arg\max_{m_n} \{V^Q(t, C_L, m_n) - V^Q(t, C_L, m_n-1) \geq p_s\}$, where $C_L(t, m_n)$ is as defined previously. The handling of an incoming long lead-time order depends on the value of $V^Q(t, n, m_n) - V^Q(t, n, m_n-1) - p_f$. If this value is negative, the seller should accept the order and reject it otherwise. On the left side of $C_L(t, m_n)$, define $C_R(t, m_n)$ as $C_R(t, m_n) = \arg\max_n \{V^Q(t, n, m_n) - V^Q(t, n, m_n-1) \geq p_f\}$. According to the

AMM, $C_R(t, m_n)$ is non-decreasing in $m_n(t)$. On the right side of $C_L(t, m_n)$, because of the subsequent upgrade decision, $C_R(t, m_n)$ becomes a line parallel to the n-axis, with a value defined as $C_R(t) = \arg\max_{m_n}\{V^Q(t, C_L, m_n) - V^Q(t, C_L, m_n - 1) \geq p_f\}$.

The three switching curves, $C_L(t, m_n)$, $C_S(t)$, and $C_R(t)$ as defined previously, jointly determine the seller's optimal inventory control policies for short and long lead-time orders, as illustrated in Fig. 4.1(a) and (b), respectively.

Theorem 4.5 *In period t, the seller's optimal inventory control policy depends on the customer arrival.*

(i) No customer: If $n(t) > C_L(t, m_n)$, the seller should upgrade outstanding long lead-time orders until the on-hand inventory level reaches $C_L(t, m_n)$.

(ii) A short customer: The optimal policy is characterized by two switching curves, $C_L(t, m_n)$ and $C_S(t)$, as depicted in Fig. 4.1(a). In Region AU (RU), the seller accepts (rejects) the short lead-time order and then upgrades the outstanding long lead-time orders until the switching curve $C_L(t, m_n)$ is reached. In Region A (R), the seller simply accepts (rejects) the short lead-time order, without any subsequent upgrade. Regions AU (A) and RU (R) are separated by the switching curve $C_S(t)$.

(iii) A long customer: The optimal policy is characterized by two switching curves, $C_L(t, m_n)$ and $C_R(t)$, as depicted in Fig. 4.1(b). In Region AU (RU), the seller accepts (rejects) the long lead-time order and then upgrades the outstanding long lead-time orders until the switching curve $C_L(t, m_n)$ is reached. In Region A (R), the seller simply accepts (rejects) the long lead-time order, without any subsequent upgrade. Regions AU (A) and RU (R) are separated by the switching curve $C_R(t)$.

Previous work has shown that the optimal inventory replenishment policy for inventory systems with a capacity constraint mostly follows a modified base-stock type (Evans 1967; Federgruen and Zipkin 1986; Decroix and Arreola-Risa 1998). That is, the inventory manager tries to restore its inventory to the base-stock level, only now the order quantity is constrained by the capacity. In the following analysis, we show that the optimal replenishment policy in our model is of a similar type.

Consider the replenishment decision at the beginning of the cycle discussed above. At the beginning of the cycle, we assume that the seller starts with an initial inventory $n(0)$. The seller then replenishes the inventory, if necessary, up to a desired level $n(1)$, subject to $n(1) \geq n(0)$. (Note that the capacity constraint in our model is imposed as a soft constraint.)

$$\begin{aligned} &\max_{n(1)\geq n(0)} \{V^Q(1, n(1), n(1)) - c[n(1) - n(0)] - c_p[n(1) - n(0) - Q]^+\} \\ &= cn(0) + \max_{n(1)\geq n(0)} \{V^Q(1, n, n) - cn(1) - c_p[n(1) - n(0) - Q]^+\}, \end{aligned}$$

where $c[n(1) - n(0)]$ represents the total procurement cost and $c_p[n(1) - n(0) - Q]^+$ represents the extra penalty cost if the capacity is exceeded. According to the preceding analysis, we know that $V^Q(1, n(1), n(1))$ satisfies the AMM.

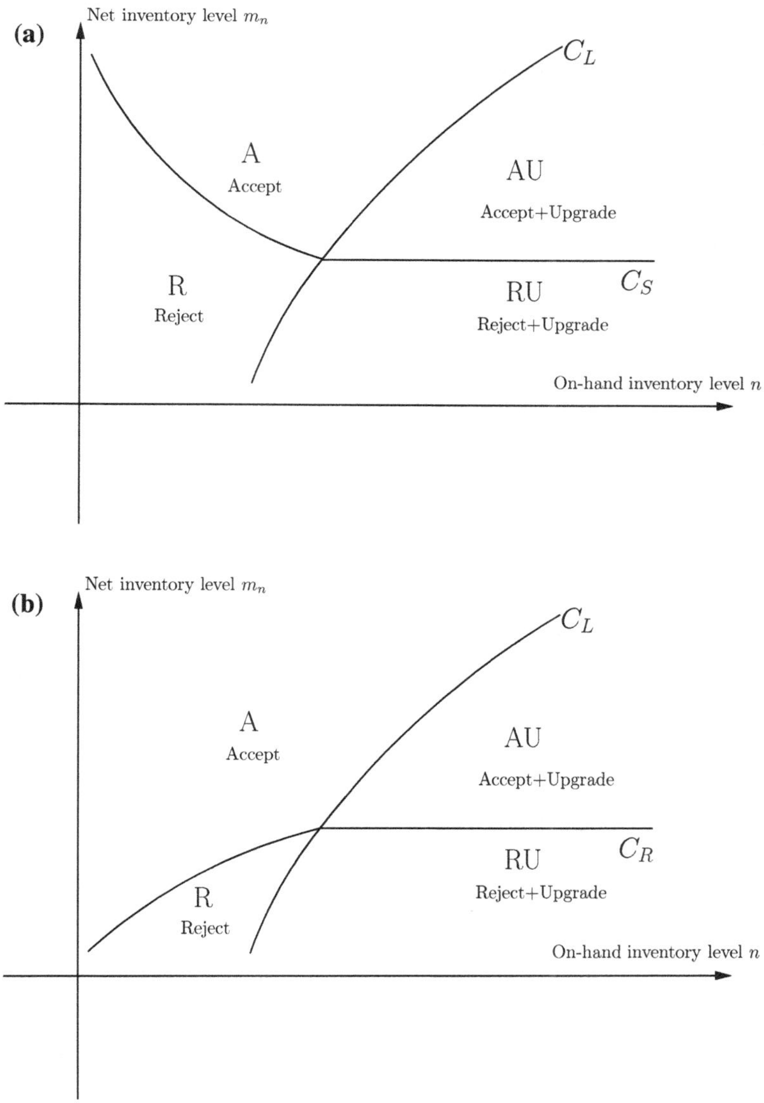

Fig. 4.1 Optimal inventory control policy: **a** for a short lead-time order; **b** for a long lead-time order

Theorem 4.6 *Let*

$$n^*(1) = \min\{n \in \mathbb{Z}^+ \mid V^Q(1, n+1, n+1) - V^Q(1, n, n) < c\},$$
$$n^{**}(1) = \min\{n \in \mathbb{Z}^+ \mid V^Q(1, n+1, n+1) - V^Q(1, n, n) < c + c_p\}.$$

The seller's optimal inventory replenishment policy is:

(i) if $n(0) \geq n^*(1)$, do not order;
(ii) if $n^*(1) - Q \leq n(0) < n^*(1)$, order up to $n^*(1)$; and
(iii) if $n(0) < n^*(1) - Q$, order up to $n^{**}(1) \vee (n(0) + Q)$.

To extend the foregoing results to the general multi-cycle case, it suffices to prove that the AMM is preserved by inventory replenishments between cycles. The sequence of events between cycles is as follows. The seller first clears all of the outstanding long lead-time orders and procures to cover the shortage, if any. The seller then replenishes its inventory, if necessary, to the optimal level. Suppose the inventory is replenished in period t, and denote the time immediately before and after the replenishment as t^- and t^+. We have

$$V^Q(t^-, n, m_n) = \max_{n^* \in \mathbb{Z}^+} \{V^Q(t^+, n^* \vee m_n, n^* \vee m_n) - c(n^* \vee m_n - m_n) - c_p(n^* \vee m_n - m_n - Q)^+\}.$$

We prove the following result.

Lemma 4.5 *The optimal inventory replenishment between cycles preserves the AMM.*

Given that the AMM holds for all $V^Q(t, n, m_n)$, we can generalize the preceding results derived for the single-cycle problem to the multi-cycle case.

Theorem 4.7 *For the multi-cycle problem, the optimal inventory control policy is the same as described in Theorem 4.5, and the optimal inventory replenishment policy at the beginning of each cycle is of a base-stock type, as described in Theorem 4.6.*

4.4 Demand and Inventory Management at Toyota China Dealerships

According to the results of our field study, Toyota China dealerships adopt a strategy resembling the short/long lead-time customer segmentation strategy. In this section, we carry out numerical studies using parameters calibrated from the empirical data obtained from Toyota dealerships. The results show that the inventory upgrade mechanism can effectively reduce inventory risk and achieve better inventory usage.

We adopt the non-capacitated model in the numerical studies. Recall that in the non-capacitated setting, the optimal inventory upgrade policy in a period is characterized by a threshold level that is independent of the net inventory. The seller upgrades down to the threshold level if the on-hand inventory level is above the threshold and makes no upgrade otherwise. The length of a cycle is set to one month (thirty days), and we assume that there are forty-eight periods per day.

The model parameters can be put into two categories: revenue-structure parameters, including p_s, p_f, c, h, and s (we assume there is no backorder cost), and arrival rates, including π_s and π_f. The revenue-structure parameters are retrieved directly

Table 4.1 Parameter values

p_s	p_f	c	h	b	s	π_s	π_f
2.2×10^5	2×10^5	1.82×10^5	26.5	0	1.78×10^5	0.1469	0.1833

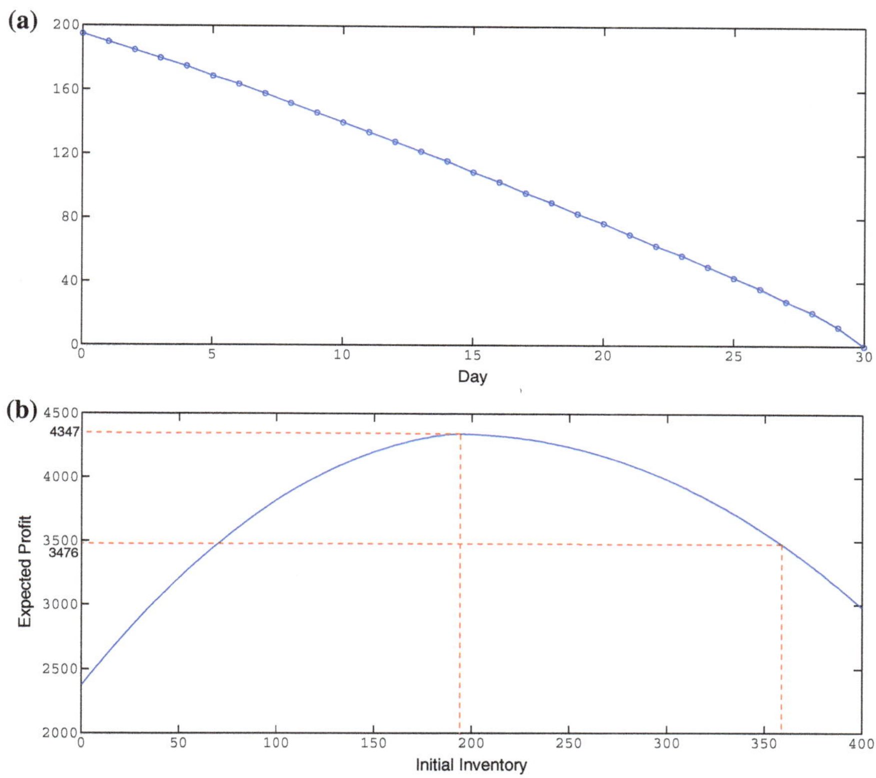

Fig. 4.2 **a** Optimal upgrade threshold and **b** Expected profit with different given initial inventory levels

from public sources, and the arrival rates are estimated from the empirical data. The parameter values are summarized in Table 4.1. Their derivation is detailed in the appendix.

For convenience, we assume that upgrades are made daily, i.e., every 48 periods, rather than in each period. Note that this is equivalent to considering the case of batch-demand arrival in a period. The optimal upgrade policy for the non-capacitated model remains of a threshold type. The optimal upgrade threshold and expected profit for given initial inventory levels are presented in Fig. 4.2. The theoretical optimal initial inventory is 194 for the upgrade model. However, the average monthly inventory level from Toyota's empirical data is roughly 360. This results in a 20 % profit difference. (Note that for easy exposition, we have scaled the price and cost parameters in the

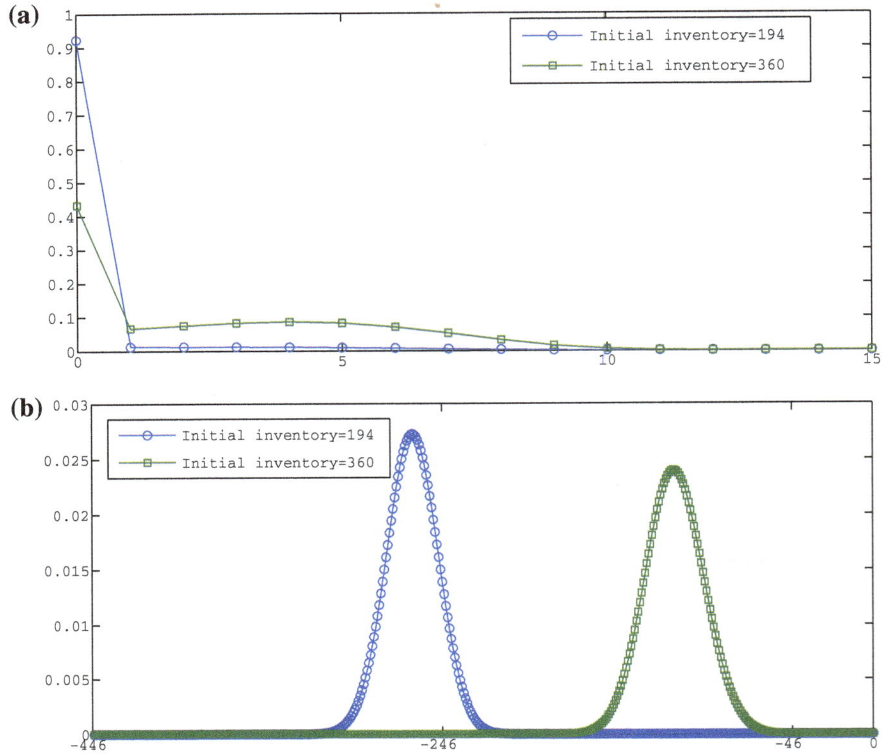

Fig. 4.3 **a** Ending on-hand inventories; **b** Ending net inventories

simulation.) The overstock may be attributed to factors that are not considered in the theoretical model. For example, there are multiple configurations for the same model. Also, Toyota dealerships may be more conservative toward demand loss than we assume (e.g., due to the long-term revenue from automobile maintenance).

We also examine the inventory usage under the two initial inventory levels by calculating the ending inventory, which partially captures how the inventory matches the demand. The ending inventory is further characterized by the ending on-hand and net inventories. The results for the ending inventory are illustrated in Fig. 4.3, and some basic statistics are provided in Table 4.2. In Fig. 4.3(a) and (b), the horizontal axis represents the inventory levels and the vertical axis represents their corresponding probability masses.

As previously mentioned, the empirical data imply that the dealership keeps an inventory level that is substantially higher than the optimal level projected by our theoretical model. The ending inventory statistics show that even if the high real-life inventory level does lead to an overstock, the upgrade mechanism can effectively ameliorate the negative effect. First, the theoretical optimal initial inventory of 194 is more likely to lead to a near-zero ending on-hand inventory (the probability mass

Table 4.2 Ending-inventory statistics

Initial inventory level	Ending on-hand inventory		Ending net inventory	
	Mean	Std	Mean	Std
360	2.47	2.75	−113	16.7
194	0.302	1.22	−263	14.6

is 0.9227), and the real-life level of 360 leads to a slightly higher ending level on average. However, the gap between the ending inventories is much smaller than the initial gap at the beginning. Without upgrades, the expected ending on-hand inventory should be roughly $360 - 0.1469 * 1.440 = 148$. The actual ending level of 2.47 is dramatically lower, which indicates that the upgrade mechanism plays an important role in the process. The same logic can be applied to the case in which the initial inventory is 194. The expected ending on-hand inventory considering only the short orders should be $194 - 148 = 46$, and the actual level is 0.3024. This also indicates that the upgrade has been used in the process. Second, although the ending on-hand inventories exhibit different patterns, the ending net inventories in the two cases are similar because, as we show in our preceding analysis, the upgrade mechanism has no effect on the evolution of the net inventory and both levels of 194 and 360 are large enough to prevent severe demand loss. These results show that the upgrade mechanism could help dealerships to effectively reduce inventory risk and achieve better inventory usage.

4.5 Concluding Remarks

In this chapter, we extend the unitary inventory control policy discussed in Chap. 3 to the general upgrade policy. The seller can choose the time to deliver the backlogged long lead-time orders. To facilitate analysis, we adopt a variable substitution to replace the outstanding long orders by the net inventory, which equals the on-hand inventory level minus the number of promised-for-delivery orders. In our analysis, we resort to the technique of anti-multimodularity to derive the structural properties of the value function. At last, we calibrate the model parameters using the empirical data collected from Toyota China dealerships and calculate the optimal policies and several performance metrics accordingly. The results show that the upgrade mechanism can improve both profit and inventory usage.

Appendix 1

This appendix contains some of the proofs. The proofs for the results in the non-capacitated case are omitted to save space because they are similar to those in Chap. 3.

Proof of Lemma 4.4

Proof We first introduce the connection between multimodularity and convexity. A function f defined on $\mathbb{Z}^m$ can be extended to a corresponding $\widetilde{f}$ defined on $\mathbb{R}^m$ by conducting linear interpolation on *atoms*. In our model, the *atoms* are in the form of $\{(n, m_n), (n-1, m_n), (n-1, m_n-1)\}$ and $\{(n, m_n), (n, m_n-1), (n-1, m_n-1)\}$. The function $\widetilde{f}$ is uniquely defined.

Theorem 4.8 *(Altman et al. 2003) f is multimodular if and only if $\widetilde{f}$ is convex.*

The following corollary is then straightforward.

Corollary 4.1 *f is anti-multimodular if and only if $\widetilde{f}$ is concave.*

We use induction to prove the lemma. First, at the end of the cycle, it is clear that the terminal value function, $V^Q(T, n, m_n)$, satisfies all three properties. Assume that the properties hold for $V^Q(t+1, n, m_n)$. We prove that they still hold for $V^Q(t, n, m_n)$.

It is obvious that the AMM is preserved under linear summation. Hence, to prove $V^Q(t, n, m_n)$ is AMM in $(-n, m_n)$, we prove that in the dynamic programming recursion all of the terms preserve the AMM, in particular the first three terms associated with the seller's inventory-control decisions. Without loss of generality, our following analysis focuses on the term $\max\{V^Q(t+1, n-1-u_s, m_n-1)+p_f, V^Q(t+1, n, m_n-1)+p_f, V^Q(t+1, n-u_0, m_n)\}$.

Define an operator $\mathscr{A}$ as follows.

$$\begin{aligned}\mathscr{A}(V^Q(t+1, n, m_n)) = \ & \max\{V^Q(t+1, n-1-u_s, m_n-1)+p_f,\\ & V^Q(t+1, n, m_n-1)+p_f, V^Q(t+1, n-u_0, m_n)\}.\end{aligned}$$

To avoid the trouble of checking all of the conditions separately, we make use of the equivalence between the AMM and continuous joint concavity. We first introduce the following notations:

$\widetilde{V}^Q$: the linear interpolation of V^Q;
$\mathscr{A}(\widetilde{V}^Q)$: the optimization on $\widetilde{V}^Q$;
$\mathscr{A}(V^Q)$: the optimization on V^Q; and
$\widetilde{\mathscr{A}}(V^Q)$: the linear interpolation of the optimized function $\mathscr{A}(V^Q)$.

Based on Corollary 4.1, if $V^Q(t+1, n, m_n)$ is AMM in $(-n, m_n)$, its linear interpolation $\widetilde{V}^Q(t+1, x, y)$ is jointly concave in $(-x, y)$. Then, we prove that after the optimization, $\mathscr{A}(\widetilde{V}^Q(t+1, x, y))$ is also jointly concave. The second step is to prove that $\mathscr{A}(\widetilde{V}^Q(t+1,x,y))$ is equivalent to $\widetilde{\mathscr{A}}(V^Q)$. This actually indicates that the order of linear interpolation and optimization can be switched without affecting the final result. In summary, the two steps are as follows:

1. $\mathscr{A}(\widetilde{V}^Q(t+1,x,y))$ is jointly concave in $(-x,y)$; and
2. $\mathscr{A}(\widetilde{V}^Q(t+1,x,y)) \equiv \widetilde{\mathscr{A}}(V^Q(t+1,n,m_n))$.

After we prove that $\widetilde{\mathscr{A}}(V^Q)$ is jointly concave, according to Corollary 4.1, $\mathscr{A}(V^Q)$ is AMM.

The operator, $\mathscr{A}$, is defined on a discrete-state space. We now extend its definition to a continuous-state space. Note that the feasible area for our problem is

$$\begin{cases} n \geq m_n, \ m_n \geq 0; \\ n \geq 0, \quad m_n < 0\,. \end{cases}$$

We divide the feasible area into atoms, and obtain $\mathscr{A}(\widetilde{V}^Q(t+1,x,y))$ by conducting linear interpolation on those atoms.

Similar to the discrete optimization, we now define $\mathscr{A}(\widetilde{V}^Q)$ as follows:

$$\mathscr{A}(\widetilde{V}^Q(t+1,x,y)) = \max_{u_y \in \mathscr{U}(x,y-\alpha), 0 \leq \alpha \leq 1} \{\widetilde{V}^Q(t+1, x-u_y, y-\alpha) + \alpha p_l\},$$

where $\mathscr{U}(x,y-\alpha) = \{u : 0 \leq u \leq x \wedge (x-y+\alpha), u \in \mathbb{R}\}$. Given that $\widetilde{V}^Q(t,x,y)$ is the piecewise linear interpolation of $V^Q(t,n,m_n)$, the continuous form of $\mathscr{A}$ subsumes the original discrete form, which is stated below.

Lemma 4.6 *At points where $x = n$, $y = m_n$ are integers,*

$$\begin{aligned} \mathscr{A}(\widetilde{V}^Q(t+1,x,y)) = \max_{\substack{u_0 \in \mathscr{U}(n,m_n), \\ u_s \in \mathscr{U}(n-1,m_n-1)}} &\{\widetilde{V}^Q(t+1, n-1-u_s, m_n-1) + p_f, \\ &\widetilde{V}^Q(t+1,n,m_n-1) + p_f, \widetilde{V}^Q(t+1, n-u_0, m_n)\}. \end{aligned}$$

Step 1:

In this part, we prove that $\mathscr{A}(\widetilde{V}^Q(t+1,x,y))$ preserves the concavity. For the continuous function, the joint concavity in (x,y) is equivalent to the joint concavity in $(-x,y)$; thus, we simply use (x,y) without loss of generality.

We first prove that $\widetilde{V}^Q(t+1, x-u_y, y-\alpha) + \alpha p_f$ is jointly concave in (x, y, u_y, α) based on the definition of joint concavity. For any two points $(x_1, y_1, u_{y,1}, \alpha_1)$ and $(x_2, y_2, u_{y,2}, \alpha_2)$, let $x_3 = \rho x_1 + (1-\rho)x_2$, $y_3 = \rho y_1 + (1-\rho)y_2$, $u_3 = \rho u_{y,1} + (1-\rho)u_{y,2}$, $\alpha_3 = \rho\alpha_1 + (1-\rho)\alpha_2$, where $0 < \rho < 1$. Given that $V^Q(t+1,x,y)$ is jointly concave in (x,y), we can prove that

$$\begin{aligned} &\widetilde{V}^Q(t+1, x_3-u_{y,3}, y_3-\alpha_3) + \alpha_3 p_f \\ \geq\ &\rho[\widetilde{V}^Q(t+1, x_1-u_{y,1}, y_1-\alpha_1) + \alpha_3 p_f] + (1-\rho)(\widetilde{V}^Q(t+1, x_2-u_{y,2}, y_2-\alpha_2) + \alpha_3 p_f). \end{aligned}$$

Also noting that $\{(x, y, u_y, \alpha) : x \geq y^+, 0 \leq u_y \leq x \wedge (x - y + \alpha), 0 \leq \alpha \leq 1\}$ is a convex set, we find that $\mathscr{A}(\widetilde{V}^Q(t+1, x, y))$ is jointly concave in (x, y).

Step 2:

We next show that $\mathscr{A}(\widetilde{V}^Q)$ is equivalent to $\widetilde{\mathscr{A}}(V^Q)$. That is, at any feasible point (x, y), the value of $\mathscr{A}(\widetilde{V}^Q)$ equals that of $\widetilde{\mathscr{A}}(V^Q)$. To prove this equivalence, we prove that the value under the operator $\mathscr{A}(\widetilde{V}^Q)$ at (x, y) within the feasible region is reachable by the operator $\widetilde{\mathscr{A}}(V^Q)$ and vice versa, i.e., $\mathscr{A}(\widetilde{V}^Q) \leq \widetilde{\mathscr{A}}(V^Q)$ and $\mathscr{A}(\widetilde{V}^Q) \geq \widetilde{\mathscr{A}}(V^Q)$. Hence, the equivalence can be established.

We first introduce the following preliminary result to facilitate the proof.

Lemma 4.7 *For the three vertices of an atom, under $\mathscr{A}(\widetilde{V}^Q)$, the points at which they take values are still vertices of a specific atom.*

We provide a sketch of the proof as follows. Denote the three vertices of an atom by (n_1, m_{n1}), (n_2, m_{n2}), and (n_3, m_{n3}). Suppose that $\mathscr{A}(\widetilde{V}^Q(t+1, n_1, m_{n1})) = \widetilde{V}^Q(t+1, n_1^*, m_{n1}^*) + (m_{n1} - m_{n1}^*)p_f$. (n_1^*, m_{n1}^*) is the point at which (n_1, m_{n1}) takes a value under $\mathscr{A}(\widetilde{V}^Q)$. Denote those for (n_2, m_{n2}) and (n_3, m_{n3}) by (n_2^*, m_{n2}^*) and (n_3^*, m_{n3}^*), respectively. Given that $\widetilde{V}^Q$ is the linear interpolation of V^Q, (n_1^*, m_{n1}^*), (n_2^*, m_{n2}^*), and (n_3^*, m_{n3}^*) are all integer points. Moreover, because V^Q is AMM, (n_1^*, m_{n1}^*), (n_2^*, m_{n2}^*), and (n_3^*, m_{n3}^*) are still vertices of the same atom.

With the aid of Lemma 4.7, we now prove the equivalence of $\mathscr{A}(\widetilde{V}^Q)$ and $\widetilde{\mathscr{A}}(V^Q)$. First, we prove that the optimal values for $\widetilde{\mathscr{A}}(V^Q)$ can be achieved by $\mathscr{A}(\widetilde{V}^Q)$. If $V^Q(t+1, n, m)$ is optimized first, for a given point (x, y) located within a specific atom, its value after the linear interpolation equals the weighted sum of the optimal values of the three vertices of that atom. According to Lemma 4.7, the optimal values of the three vertices are the values of the vertices of a specific atom. Hence, the value at (x, y) can be reproduced in that specific atom.

Next, we prove that the reverse also holds. Consider $\mathscr{A}(\widetilde{V}^Q)$. For a point (x, y) within an atom, referred to as Φ, every point within its optimization domain can be expressed as the weighted sum of the values of the vertices of some atom (the shift parameter, αp_f, does not affect this result). Also note that those vertices fall within the respective optimization domain of the three vertices of Φ. This implies that $\mathscr{A}(\widetilde{V}^Q) \leq \widetilde{\mathscr{A}}(V^Q)$.

Given the results from Steps 1 and 2, we know that $\mathscr{A}(V^Q(t+1, n, m_n))$ is AMM in $(-n, m_n)$.

We now look at the first two terms in the dynamic programming recursion. The proof of the AMM of $\max V^Q(t+1, n - u_0, m_n)$ is straightforward because the optimization only involves n. Regarding $\max\{V^Q(t+1, n-1-u_s, m_n - 1) + p_s, V^Q(t+1, n - u_0, m_n)\}$, using a logic similar to that in the proof of the first term, we define an operator $\mathscr{B}$ as

$$\mathscr{B}(\widetilde{V}^Q(t+1, x, y)) = \max_{u_y \in \mathscr{U}(x, y-\alpha), 0 \leq \alpha \leq 1} \{\widetilde{V}^Q(t+1, x - u_y, y - \alpha) + \alpha p_s\}.$$

The rest of the proof then follows that of the AMM of $\max\{V^Q(t+1, n-1-u_s, m_n - 1) + p_f, V^Q(t+1, n, m_n - 1) + p_f, V^Q(t+1, n - u_0, m_n)\}$.

Proof of Theorem 4.6

Proof First, let $\triangle V^Q(1, n, n) = V^Q(1, n+1, n+1) - V^Q(1, n, n)$. Define $n^*(1)$ as the smallest n that satisfies $\triangle V^Q(1, n, n) < c$ and $n^{**}(1)$ as the smallest n that satisfies $\triangle V^Q(1, n, n) < c + c_p$. From the AMM property of $V^Q(1, n, n)$, we have $n^*(1) \geq n^{**}(1)$ and $V^Q(0, n^*, n^*) \geq V^Q(0, n^{**}, n^{**})$.

$n^*(1)$ can be considered as the base-stock level without capacity constraint. It is optimal for the seller to order up to $n^*(1)$ if it is within the capacity limit, and not to order if $n(0) \geq n^*(1)$. The question is how to replenish when $n^*(1)$ exceeds the capacity limit. We list all of the possibilities as follows:

1. if $n(0) \geq n^*(1)$, it is optimal not to order;
2. if $n^*(1) - Q \leq n(0) < n^*(1)$, it is optimal to order up to $n^*(1)$; and
3. if $n(0) < n^*(1) - Q$, it is optimal to order up to $n^{**}(1) \vee (n(0) + Q)$.

Appendix 2: The Derivation of Parameter Values

This appendix contains the derivation of the parameter values used in the numerical studies in Sect. 4.4. As mentioned in Sect. 4.4, the model parameters can be put into two categories: those related to the revenue structure, including p_s, p_f, c, h, and s, and the customer arrival rates, including π_s and π_f. The revenue-structure parameters are retrieved directly from public sources. The two prices, p_f and p_s, are simply considered as the manufacturer's suggested retail price (MSRP) and the marked-up price for spot sales, respectively. The mark-up degree is about 10 % of the MSRP. We use the common operating margin and p_f to derive c. The common operating margin is 10 % so c is set to 91. For s, rather than the true salvage value, we refer to the depreciated value of the car. As the duration of a cycle is set as one month (which will be explained later), s is set to 98 % of c. We assume that there is no backorder cost. The inventory holding cost h is considered to be the financing cost rather than the literal "holding cost." Automakers usually require instant payment from dealerships upon placing orders, but dealerships do not receive full payment from customers until delivery. Although we are aware of other operating expenses, we consider this financial cost as the main source of h. We use the benchmark one-year lending rate issued by the People's Bank of China, which was 5.31 % at the time of writing, as the loan interest rate. Hence, h is calculated as the daily interest. The parameters related to the revenue structure are summarized in Table 4.3. All of the values are denominated in Chinese yuan (CNY).

Table 4.3 Parameters related to the revenue structure

p_s	p_f	c	h	b	s
2.2×10^5	2×10^5	1.82×10^5	26.5	0	1.78×10^5

Maximum Likelihood Estimation of Arrival Rates from the Empirical Data

Customer arrival rates are not directly retrievable and must be estimated from the empirical data. As detailed in Chap. 2, over the years, the dealerships have developed a comprehensive customer relationship management tool known as e-CRB. i-CROP, a component of e-CRB, classifies potential customers into four classes (A, B, C, and D) based on their desired delivery times. The delivery times are one week, one month, three months, and more than three months after the purchasing time, respectively. We consider this classification as a good fit for our lead-time differentiated model. We collected the daily classification data in June and July 2009 by recording the daily arrivals of class A, B, C, and D customers at a dealership requesting a specific automobile model. Toyota operates a moving-window type of order planning practice with its dealerships. Dealerships are required to provide order numbers three months before the delivery date. Two months before the delivery, dealerships have the opportunity to adjust the order quantities previously placed within a certain range. One month before the delivery, Toyota fixes the order quantity but allows dealerships to make changes to the color. In this setting, we consider class A and B customers as short lead-time customers, and class C and D customers as long lead-time customers. This data categorization also comes from the observation that dealerships demand a premium for delivery within one or two months. The length of a cycle is set to one month (thirty days).

Sixty samples were collected. After analyzing the data and considering the unitary demand assumption in our model, we assume that there are 48 periods per day. We use the maximum likelihood estimation (MLE) method to derive π_s and π_f. Given that there are k days,

$$\pi_s = \frac{\sum_{i=1}^{k} y_{si}}{48k}, \pi_l = \frac{\sum_{i=1}^{k} y_{fi}}{48k},$$

where y_{si} and y_{fi} represent the daily numbers of short and long lead-time customers in day i. From the data, we can obtain the values of π_s and π_f: $\pi_s = 0.1469$, and $\pi_f = 0.1833$.

References

E. Altman, B. Gaujal, and A. Hordijk. *Discrete-Event Control of Stochastic Networks: Multimodularity and Regualrity* (Springer, 2003)

G. Decroix, A. Arreola-Risa, Optimal production and inventory policy for multiple products under resrouce constraints. Manage. Sci. **44**(7), 950–961 (1998)

R.V. Evans, Inventory control of a multiproduct system with a limited production resource. Naval Res. Logist. Q. **14**(2), 173–184 (1967)

A. Federgruen, P. Zipkin, An inventory model with limited production capacity and uncertain demands II. The discounted-cost criterion. Math. Oper. Res. **11**(2), 208–215 (1986)
B. Hajek, Extreme splittings of point processes. Math. Oper. Res. **10**(4), 543–556 (1985)
W. Narongwanich, I. Duenyas, J. Birge, Optimal portfolio of reconfigurable and dedicated capacity under uncertainty. Working Paper, Northwestern University, 2002

Chapter 5
Inventory Control and Pricing with Alternative Delivery Times

Abstract In this chapter, we consider joint inventory and pricing decisions. The seller can adjust the price to control the arrival rates directly. We derive the optimal joint policies and then compare the two mechanisms using a hybrid model in which the seller must choose between the two control means in each period. Our results indicate that the inventory decision is meaningful in the high-inventory case, and the pricing decision is advantageous when the inventory level is low. We further demonstrate the equivalence of the two models and show that the optimal policy in the hybrid model is a switching policy characterized by a switching threshold.

5.1 Introduction

In the previous two chapters, we discuss the inventory control policies for inventory models with two alternative delivery lead times. In those models, the seller maximizes its profit through inventory decisions, such as whether to upgrade outstanding long lead-time orders and reject incoming orders. The prices for the delivery options are assumed given. In this chapter, we consider the possibility of price adjustment and study the optimal joint pricing and inventory control policies.

The incorporation of the pricing decision is of interest and value for several reasons. First, the fixed-price assumption is appropriate when prices are determined by the market and relatively rigid. However, there are also cases in which firms are able to adjust the prices to reflect the real-time demand-supply situation. Note that the price adjustment can be realized in a variety of forms such as through personalized coupons or bundling. For example, in the Toyota dealership example introduced in Chap. 2, dealerships may ask those customers who choose spot delivery to buy a bundled insurance product or upgrade the configuration of the car. The study of the optimal joint policies thus can help with firms' decision making. Second, as we show in later analysis, in the presence of the delivery-lead-time-based customer segmentation, inventory and pricing decisions have similarities and differences. Both means aim to maximize firms' profit, but each has its own merit. The inventory decision can mimic the pricing decision through a probabilistic upgrade policy in certain circumstances, and the pricing decision affects the customer arrival rates and revenue

X. Liang et al., *Inventory Management with Alternative Delivery Times*,
SpringerBriefs in Operations Management, DOI 10.1007/978-3-319-48635-2_5

directly. To better illustrate these points, we consider the two mechanisms in two different scenarios. In the first scenario, the seller makes inventory and pricing decisions jointly, and in the second scenario, the seller uses only one of the two means in each period. An interesting finding is that the two scenarios are actually equivalent in certain model settings.

In the remainder of this chapter, Sect. 5.2 presents the formulation of the joint-decision problem. In Sect. 5.3, we obtain the optimal joint inventory and pricing decisions. In Sect. 5.4, we consider another scenario in which the seller can choose only one control means and show its equivalence to the joint-decision model. Section 5.5 concludes the chapter.

5.2 Model Formulation

The model setup still follows that of the non-capacitated upgrade model in Chap. 4, only now we incorporate the pricing decision into the model. We assume the seller can adjust the price of the short delivery lead-time option, i.e., p_s. The price is announced at the beginning of each period and determines the demand rates, π_s and π_f, in that period. We assume that customers are myopic, i.e., they are neither backward- nor forward-looking.

To facilitate analysis, we make several simplifying assumptions. We assume that the adjustment of p_s only induces demand cannibalization. That is,

$$\frac{\partial \pi_s}{\partial p_s} \leq 0, \ \frac{\partial \pi_f}{\partial p_s} \geq 0, \ \text{and} \ \frac{\partial \pi_0}{\partial p_s} = 0. \tag{5.1}$$

The first two inequalities are intuitive. The last equation indicates that the price adjustment does not incur any demand gain/loss. For example, in the Toyota dealership example in Chap. 2, dealerships can adjust the price premium they charge for the spot-delivery option. As the premium increases, more customers are likely to switch from the spot option to the consignment option, but they buy anyway.

Note that there is a monotonic mapping between p_s and π_s. We thus can use π_s to replace p_s for analytical convenience. The inequalities in (5.1) can be rewritten as follows:

$$\frac{\partial p_s}{\partial \pi_s} \leq 0, \ \frac{\partial \pi_f}{\partial \pi_s} \leq 0, \ \text{and} \ \frac{\partial \pi_0}{\partial \pi_s} = 0. \tag{5.2}$$

Also, given that $\pi_s + \pi_f + \pi_0 = 1$, we know that

$$1 + \frac{\partial \pi_f}{\partial \pi_s} = 0, \ \text{and} \ \frac{\partial^2 \pi_f}{\partial \pi_s^2} = 0.$$

Another assumption is made for the one-period revenue function $\pi_s p_s + \pi_f p_f$. We assume that it is twice differentiable and concave in π_s. That is,

$$\frac{\partial^2(\pi_f p_f + \pi_s p_s)}{\partial \pi_s^2} \leq 0.$$

This assumption is consistent with the dynamic pricing literature. It is easy to verify that the concavity holds for common demand forms such as linear, multinomial logit (MNL), and Cobb-Douglas models.

Denote the feasible range of π_s as $[\underline{\pi_s}, \overline{\pi_s}]$. Correspondingly, the range of π_f is $[\underline{\pi_f}, \overline{\pi_f}]$. The dynamic programming equations can be written as follows:

$$\begin{aligned} V(t,n,m_n) = &\max_{\pi_s \in [\underline{\pi_s}, \overline{\pi_s}]} \{\pi_s p_s + \pi_s \max_{u_s \in \mathscr{U}(n-1,m_n-1)} V(t+1, n-1-u_s, m_n-1) \\ &+ \pi_f p_f + \pi_f \max_{u_f \in \mathscr{U}(n,m_n-1)} V(t+1, n-u_f, m_n-1)\} \\ &+ \pi_0 \max_{u_0 \in \mathscr{U}(n,m_n)} V(t+1, n-u_0, m_n) - hn, \\ V(t,0,m_n) = &(\overline{\pi_0} + \underline{\pi_s}) V(t+1,0,m_n) + \overline{\pi_f}[p_f + V(t+1,0,m_n-1)], \\ V(T,n,m_n) = &-cm_n^- + sm_n^+. \end{aligned} \tag{5.3}$$

Note that p_s and π_f are both functions of π_s. It is obvious that when $n(t) = 0$, the optimal policy is to set π_s to $\underline{\pi_s}$ to induce as many customers as possible to choose the long lead-time option.

5.3 Optimal Inventory Control and Pricing Policy

We start with a single-cycle problem and deduce the following structural properties of the value function.

Lemma 5.1 *1. $V(t,n,m_n)$ is integer concave in n;*
2. $V(t,n,m_n)$ is modular in n and m_n.

Lemma 5.1 shows that the value function still exhibits the same properties as in an inventory-control-only model. Based on the structural properties, the optimal joint pricing and upgrade policies can be obtained as follows.

Theorem 5.1 *In each period t, there exist an optimal price p_s^* and a threshold level $C_L(t)$. It is optimal for the seller to upgrade the inventory level down to $C_L(t)$ if $n(t) > C_L(t)$ and set p_s to p_s^*, which is a function of $n(t) \vee C_L(t)$.*

Theorem 5.1 indicates that the optimal price is not fixed but rather contingent on the inventory level. Furthermore, we can prove the following monotonic relationship between p_s^* and $n(t)$.

Corollary 5.1 *In period t, the optimal price p_s^* is non-increasing in $n(t)$.*

The derivation of the optimal inventory replenishment policy is the same as that in Chap. 4 because no pricing decision is involved. An additional property is necessary.

Lemma 5.2 *$V(t, n, m_n)$ is integer concave in $m_n(t)$.*

At the beginning of the cycle, we assume the seller starts with an initial inventory $n(0)$. The inventory replenishment decision can be expressed by the following equation:

$$\begin{aligned}&\max_{n(1)\geq n(0)} \{V(1, n(1), n(1)) - c[n(1) - n(0)]\}\\ =\ & cn(0) + \max_{n(1)\geq n(0)} \{V(1, n(1), n(1)) - cn(1)\}.\end{aligned}$$

Given the structural properties in Lemmas 5.1 and 5.2, we can derive the optimal inventory replenishment policy.

Theorem 5.2 *The optimal inventory replenishment at the beginning of the cycle follows a base-stock type.*

The extension to the multi-cycle problem is straightforward.

Lemma 5.3 *In period t, $V(t, n, m_n)$ satisfies the following properties.*

Integer concavity in n: $V(t, n+1, m_n) - V(t, n, m_n) \leq V(t, n, m_n) - V(t, n-1, m_n)$;
Integer concavity in m_n: $V(t, n, m_n+1) - V(t, n, m_n) \leq V(t, n, m_n) - V(t, n, m_n-1)$;
Modularity in n and m_n: $V(t, n+1, m_n) - V(t, n, m_n) = V(t, n+1, m_n+1) - V(t, n, m_n+1)$.

Theorem 5.3 *For the multi-cycle problem, the optimal upgrade policy in a period is the same as that described in Theorem 5.1, and the optimal inventory replenishment at the beginning of each cycle follows a base-stock type.*

5.4 Comparison of Inventory Control and Dynamic Pricing

In the previous section, we derive the optimal joint inventory control and pricing policies. The inventory decision in our model is realized through a delivery upgrade, and the pricing decision can be used to directly affect the arrival rates. It is also interesting to compare the two mechanisms and study their similarities and differences. For example, in Maglaras and Meissner (2006), they try to find a unified framework to study these two types of problems. In their setting, the inventory control involves choosing a mixed strategy for menu selection. They derive a common structure for the two problems and compare their performances numerically. In Lin et al. (2008), they study both capacity allocation and dynamic pricing using the framework of stochastic knapsack problems. The optimal policy is a *switch-over* policy. For capacity allocation, the prices are assumed to be given exogenously and the decision is to choose the timing when low-price orders are supplied. Using the same approach, they discuss the pricing model, which can be viewed as an optimal markdown process.

To compare the two mechanisms, we consider a model in which the seller must choose between inventory control and dynamic pricing in each period. We call this model the *hybrid* model, as opposed to the *joint* model discussed in preceding analysis. The hybrid model can shed light on the conditions under which one mechanism is preferred. For analytical convenience, we make one change to the upgrade policy in the hybrid model: we assume that the upgrade is probabilistic. That is, there is a probability associated with each feasible upgrade quantity. Note that the change to the probabilistic upgrade does not affect our previous results, but make inventory control and dynamic pricing comparable.

5.4.1 The Hybrid Model

In the hybrid model, the seller chooses either dynamic pricing or inventory control in each period. In the case of inventory control, we assume that the seller chooses to set π_s (p_s) to π_s^* (p_s^*), the value that maximizes the one-period revenue. Correspondingly, the other two arrival rates are denoted as π_f^* and π_0^*, respectively. The dynamic programming recursion for the hybrid model can be written as follows:

$$
\begin{aligned}
V(t,n,m_n) &= \max\{V^{DP}(t,n,m_n), V^{UP}(t,n,m_n)\},\\
V^{DP}(t,n,m_n) &= \max_{\pi_s\in[\underline{\pi_s},\overline{\pi_s}]} \{\pi_s p_s + \pi_s V(t+1,n-1,m_n-1)\\
&\quad +\pi_f p_f + \pi_f V(t+1,n,m_n-1)\}\\
&\quad +\pi_0 V(t+1,n,m_n) - hn,\\
V^{DP}(t,0,m_n) &= (\overline{\pi_0} + \underline{\pi_s})V(t+1,0,m_n)\\
&\quad +\overline{\pi_f}[p_f + V(t+1,0,m_n-1)],\\
V^{UP}(t,n,m_n) &= \pi_0^* \max_{\Lambda_0} \sum_{u_0\in\mathscr{U}(n,m_n)} \lambda_0(u_0)V(t+1,n-u_0,m_n)\\
&\quad +\pi_f^* \max_{\Lambda_f} \sum_{u_f\in\mathscr{U}(n,m_n-1)} \lambda_f(u_f)V(t+1,n-u_f,m_n-1)\\
&\quad +\pi_s^* \max_{\Lambda_s} \sum_{u_s\in\mathscr{U}(n-1,m_n-1)} \lambda_s(u_s)V(t+1,n-1-u_s,m_n-1)\\
&\quad +\pi_s^* p_s^* + \pi_f^* p_f - hn,\\
V^{UP}(t,0,m_n) &= (\pi_0^* + \pi_s^*)V(t+1,0,m_n)\\
&\quad +\pi_f^*[p_f + V(t+1,0,m_n-1)],\\
V(T,n,m_n) &= -cm_n^- + sm_n^+.
\end{aligned}
$$

In the preceding equations, Λ_0, Λ_f, and Λ_s are the probabilistic-upgrade vectors defined as following:

$$\Lambda_0 = \{\lambda_0(i) : i \in \mathscr{U}(n, m_n)\},$$
$$\Lambda_f = \{\lambda_f(i) : i \in \mathscr{U}(n, m_n - 1)\},$$
$$\Lambda_s = \{\lambda_s(i) : i \in \mathscr{U}(n - 1, m_n - 1)\}.$$

We compare inventory control and dynamic pricing by comparing $V^{DP}(t, n, m_n)$ and $V^{UP}(t, n, m_n)$. We look at all possible cases.

(a) $n(t) > 0$.

(i) $V(t+1, n, m_n - 1) \leq V(t+1, n-1, m_n - 1)$: for dynamic pricing, the seller should choose a $\pi_s \leq \pi_s^*$. This dynamic pricing policy can be easily replicated by the probabilistic upgrade policy. Furthermore, given that the one-period revenue stays at the optimum under upgrade, the upgrade policy outperforms the dynamic pricing. Hence, we have $V^{UP}(t, n, m_n) \geq V^{DP}(t, n, m_n)$.

(ii) $V(t+1, n, m-1) > V(t+1, n-1, m-1)$: for dynamic pricing, the seller should choose a $\pi_s > \pi_s^*$. The probabilistic upgrade policy cannot replicate the dynamic pricing policy. However, we can derive an upper bound on the performance gap. Let $\overline{p_s}$ be the price that corresponds to $\underline{\pi_s}$.

$$\begin{aligned}
& V^{UP}(t, n, m_n) - V^{DP}(t, n, m_n) \\
\geq\ & -(\pi_s^* - \underline{\pi_s})[V(t+1, n, m_n - 1) - V(t+1, n-1, m_n - 1)] \\
\geq\ & -(\pi_s^* - \underline{\pi_s})(\overline{p_s} - h).
\end{aligned}$$

The first inequality comes directly from the dynamic programming equations, and the second inequality is based on sample path analysis.

(b) $n_t = 0$.

There is no decision left for inventory control right now. For dynamic pricing, the seller should choose π_s to make π_f as high as possible. Hence,

$$\begin{aligned}
-(\overline{\pi_f} - \pi_f^*)(p_f - s) \leq\ & V^{UP}(t, 0, m_n) - V^{DP}(t, 0, m_n) \\
\leq\ & -(\overline{\pi_f} - \pi_f^*)(p_f - c).
\end{aligned}$$

Preceding analysis sheds some light on the advantage of each mechanism. When the inventory level is high, the upgrade mechanism can be more effective because it can reduce the inventory without affecting the one-period revenue. However, when the inventory level becomes low, the upgrade mechanism gradually falls out of favor. The dynamic pricing mechanism can be used to induce customers to choose the long lead-time delivery option and thus reduce the demand loss.

5.4.2 The Equivalence of the Hybrid and Joint Models

We devise the hybrid model to compare the pricing and inventory controls. It is interesting to note that in the current model setting, the hybrid model is actually equivalent to the joint model.

Theorem 5.4 *The joint and hybrid models are equivalent.*

The equivalence of the two models depends on our specific model setting and does not hold generally. In general, it is obvious that the joint model should outperform the hybrid model. The following result is straightforward given the equivalence of the two models.

Corollary 5.2 *For the hybrid model, in each period t, there exists a threshold $C_L(t)$. The seller's optimal policy is to use dynamic pricing when $n(t) \leq C_L(t)$ and to use inventory control otherwise.*

The optimal policy for the hybrid model in Corollary 5.2 demonstrates the complementarity between inventory control and dynamic pricing.

5.5 Concluding Remarks

In this chapter, we incorporate the pricing decision into the model and consider the optimal joint pricing and inventory control policies. We also compare the two mechanisms through an intermediate hybrid model and discuss the circumstances under which each mechanism outperforms. We also demonstrate that the hybrid model is equivalent to the joint model. This equivalence can be used to derive the optimal policy for the hybrid model. The optimal switching policy is of a threshold type and well justifies the complementarity of the two mechanisms.

Our theoretical analysis and results can be of practical value to real-world businesses. As indicated by the results, the lead-time-based market segmentation can complement rather than be dominated by the dynamic pricing. Backlogged long lead-time orders give firms extra flexibility when they need to reduce their inventories, and this is achieved without altering the prices.

Appendix

Proof of Lemma 5.1

Proof We prove this result by induction. It is obvious that the terminal function $V(T, n, m)$ satisfies both properties. Assuming that the two properties hold for $V(t+1, n, m)$, we prove that they also hold for $V(t, n, m)$.

Rewrite the dynamic programming equation as follows:

$$\begin{aligned}
V(t,n,m_n) &= \max_{\pi_s \in [\underline{\pi_s},\overline{\pi_s}]} \{\pi_s p_s + \pi_s \max_{u_s \in \mathscr{U}(n-1,m_n-1)} V(t+1,n-1-u_s,m_n-1) \\
&\quad + \pi_f p_f + (1-\pi_0-\pi_s) \max_{u_f \in \mathscr{U}(n,m_n-1)} V(t+1,n-u_f,m_n-1)\} \\
&\quad + \pi_0 \max_{u_0 \in \mathscr{U}(n,m_n)} V(t+1,n-u_0,m_n) - hn \\
&= \max_{\pi_s \in [\underline{\pi_s},\overline{\pi_s}]} \{\pi_s p_s + \pi_f p_f \\
&\quad + \pi_s [\max_{u_s \in \mathscr{U}(n-1,m_n-1)} V(t+1,n-1-u_s,m_n-1) \\
&\quad - \max_{u_f \in \mathscr{U}(n,m_n-1)} V(t+1,n-u_f,m_n-1)]\} \\
&\quad + (1-\pi_0) \max_{u_f \in \mathscr{U}(n,m_n-1)} V(t+1,n-u_f,m_n-1) \\
&\quad + \pi_0 \max_{u_0 \in \mathscr{U}(n,m_n)} V(t+1,n-u_0,m_n) - hn.
\end{aligned}$$

The optimal π_s depends on

$$\max_{u_s \in \mathscr{U}(n-1,m_n-1)} V(t+1,n-1-u_s,m_n-1) - \max_{u_f \in \mathscr{U}(n,m_n-1)} V(t+1,n-u_f,m_n-1),$$

and the one-period revenue function. It equals

$$\begin{aligned}
\arg_{\pi_s} \ &\{\frac{\partial(\pi_s p_s + \pi_f p_f)}{\partial \pi_s} + \max_{u_s \in \mathscr{U}(n-1,m_n-1)} V(t+1,n-1-u_s,m_n-1) \\
&- \max_{u_f \in \mathscr{U}(n,m_n-1)} V(t+1,n-u_f,m_n-1) = 0\},
\end{aligned}$$

if this value falls within $[\underline{\pi_s},\overline{\pi_s}]$; otherwise it is set to $\underline{\pi_s}$ or $\overline{\pi_s}$. Given the integer concavity of $V(t+1,n,m_n)$ in n, if $\pi_s p_s + \pi_f p_f$ is strictly concave in π_s, then for given values of $n(t)$ and $m_n(t)$, the optimal π_s is uniquely determined. Based on the integer concavity of $V(t+1,n,m_n)$ in n, we have the following result.

Lemma 5.4 *For a given m_n, $\max_{u_s \in \mathscr{U}(n-1,m_n-1)} V(t+1,n-1-u_s,m_n-1) - \max_{u_f \in \mathscr{U}(n,m_n-1)} V(t+1,n-u_f,m_n-1)$ is non-decreasing in n.*

We discuss two possible subcases as follows.

Subcase 1: given the concavity of $V(t+1,n,m_n)$, when $n(t)$ is larger than the optimal on-hand inventory level for $m_n(t)-1$, π_s is set to the value of $\arg\max_{\pi_s}(\pi_s p_s + \pi_f p_f)$ to maximize the one-period revenue.

Subcase 2: if $n(t)$ equals or is smaller than the optimal on-hand inventory level for $m(t)-1$, we know the following:

$$\max_{u_s \in \mathscr{U}(n-1,m_n-1)} V(t+1,n-1-u_s,m_n-1) = V(t+1,n-1,m_n-1),$$

$$\max_{u_f \in \mathscr{U}(n,m_n-1)} V(t+1,n-u_f,m_n-1) = V(t+1,n,m_n-1).$$

Let $\widetilde{V}(t+1, x, m_n)$ represent the continuous counterpart of $V(t+1, n, m_n)$ by linear interpolation. For $\forall x \in \mathbb{R}^+$,

$$\widetilde{V}(t+1, x, m_n) = (\lceil x \rceil - x)V(t, \lfloor x \rfloor, m_n) + (x - \lfloor x \rfloor)V(t, \lceil x \rceil, m_n),$$

where $\lceil x \rceil$ is the smallest integer that is larger than x and $\lfloor x \rfloor$ is the largest integer that is smaller than x. $\lceil x \rceil = \lfloor x \rfloor + 1$. Given the integer concavity of $V(t+1, n, m_n)$ in n, we can obtain the following two results.

Lemma 5.5 *$\widetilde{V}(t+1, x, m_n)$ is concave in x.*

Lemma 5.6 *There exists a value of $x \in [0, 1]$, such that $(1-\pi_0)\widetilde{V}(t+1, n-1+x, m_n) = \pi_s V(t+1, n-1, m_n-1) + (1-\pi_0-\pi_s)V(t+1, n, m_n-1)$. Furthermore, $x = 1-\pi_s$.*

Let

$$\phi(\pi_s) = \begin{cases} \pi_s p_s + \pi_f p_f, & \text{if } \pi_s \in [\underline{\pi_s}, \overline{\pi_s}]; \\ 0, & \text{otherwise.} \end{cases}$$

We can rewrite the value function as follows:

$$\begin{aligned} V(t, n, m_n) = & \max_{x \in [\underline{\pi_s}, \overline{\pi_s}]} [\phi(x) + (1-\pi_0)\widetilde{V}(t+1, n-x, m_n)] \\ & + \pi_0 \max_{u_0 \in \mathscr{U}(n, m_n)} V(t+1, n-u_0, m_n) - hn. \end{aligned}$$

We now prove that $\max_{x \in [\underline{\pi_s}, \overline{\pi_s}]} [\phi(x) + (1-\pi_0)\widetilde{V}(t+1, n-x, m_n)]$ preserves the integer concavity in n. Denote the optimal x for $V(t, n-1, m_n)$, $V(t, n, m_n)$, and $V(t, n+1, m_n)$ as x^*, $x^{*\prime}$, and $x^{*\prime\prime}$, respectively.

$$\begin{aligned} & \{[\phi(x^*) + (1-\pi_0)\widetilde{V}(t+1, n-1-x^*, m_n)] + [\phi(x^{*\prime\prime}) \\ & +(1-\pi_0)\widetilde{V}(t+1, n+1-x^{*\prime\prime}, m_n)]\}/2 \\ \leq\ & \phi(\frac{x^* + x^{*\prime\prime}}{2}) + (1-\pi_0)\widetilde{V}(t+1, n - \frac{x^* + x^{*\prime\prime}}{2}, m_n)] \\ \leq\ & \phi(x^{*\prime}) + (1-\pi_0)\widetilde{V}(t+1, n-x^{*\prime}, m_n). \end{aligned}$$

It is easy to see that the boundary condition at $n(t) = 0$ and the conjunction at the optimum also satisfy the integer concavity. Hence, the integer concavity in $n(t)$ holds for $V(t, n, m_n)$.

We skip the proof of modularity as it is similar to the proof of modularity in the non-capacitated upgrade model.

Proof of Corollary 5.1

Proof Given the integer concavity of $V(t, n, m_n)$ in $n(t)$, the first order condition can be written as follows:

$$\begin{cases} \dfrac{\partial(\pi_f p_f + \pi_s p_s)}{\partial \pi_s}, \\ \text{when } n(t) \text{ is larger than the optimum for } m_n(t) - 1; \\ \dfrac{\partial(\pi_f p_f + \pi_s p_s)}{\partial \pi_s} + V(t+1, n, m_n - 1) \\ \qquad\qquad -V(t+1, n-1, m_n - 1), \\ \text{otherwise.} \end{cases}$$

Hence, π_s^* is non-decreasing in $n(t)$. Correspondingly, p_s^* is non-increasing in $n(t)$.

Proof of Lemma 5.2

Proof We use induction to prove this result. First, it is easy to see that the terminal function satisfies the integer concavity. Assume that the concavity holds for $V(t + 1, n, m_n)$. We prove that it holds for $V(t, n, m_n)$ too.

From previous analysis, we already see that the pricing and upgrade controls depend on $n(t)$, regardless of $m_n(t)$. The integer concavity thus can be carried over.

Proof of Theorem 5.4

Proof First, it is obvious that the joint model dominates the hybrid model, as every strategy in the hybrid model can be replicated by the joint model. We now show that the optimal policy in the joint model can be achieved by the hybrid model. According to Theorem 5.1, when the inventory control, i.e., upgrade, is used, the short lead-time customer arrival rate is set as π_s^*. When there is no upgrade, the optimal rate is set according to the inventory level. The former case can be considered as pure inventory control, and the latter case corresponds to pure dynamic pricing. Hence, the optimal policy of the joint model is achievable by the hybrid model.

References

G. Lin, Y. Lu, D. Yao, The stochastic knapsack revisited: switch-over policies and dynamic pricing. Oper. Res. **56**(4), 945–957 (2008)

C. Maglaras, J. Meissner, Dynamic pricing strategies for multiproduct revenue management problems. Manuf. Serv. Oper. Manag. **8**(2), 136–148 (2006)

Chapter 6
Inventory Commitment and Prioritized Backlogging Clearance with Alternative Delivery Times

Abstract In this chapter, we consider prioritized backlogging clearance. Upon the arrival of a replenishment, the seller clears backlogged short and long lead-time orders according to a certain priority rule. Three priority rules are discussed: first-come first-served, myopic short first-served, and strategic short first-served. We characterize the optimal inventory commitment policies under all three priority rules and then compare their performances to shed light on the suitable conditions to apply these rules.

6.1 Introduction

Our analysis of inventory models with alternative delivery lead times so far has assumed that all of the backlogged long lead-time orders can be cleared at the beginning of the next cycle by either using replenished inventory or incurring a penalty cost if a supply capacity is exceeded. It is clear that there are exceptions to this assumption, and these exceptions give rise to new questions that must be answered. For example, in the Toyota dealership example discussed in Chap. 2, Toyota China usually rolls out a tentative production plan for the coming year at the end of the previous year, and dealerships need to submit their orders months before the delivery. All of these factors may make the inventory replenishments rigid rather than flexible as we assume in preceding analysis. Dealerships thus must deal with the problem of how to make use of these predetermined inventory replenishments to clear backlogged orders. The straightforward and intuitive way is to follow the first-come first-served (FCFS) rule; however, a priority allocation rule that shifts toward high-margin but less patient customers may be more appropriate. Through our private communications with 4S store managers, we found that dealerships usually keep a certain level of inventory for impatient customers who are willing to pay a premium for spot pickup. At the same time, to retain delayed orders, incentives such as service plans and insurance coverages are frequently offered. The same kind of prioritized backlogging clearance is also prominent in online retailing. Orders with different delivery times overlap with each other temporally. Delivering the orders simply based on their placement time may not be the best strategy.

X. Liang et al., *Inventory Management with Alternative Delivery Times*,
SpringerBriefs in Operations Management, DOI 10.1007/978-3-319-48635-2_6

In this chapter, we focus on the problem of prioritized backlogging clearance and evaluate several priority rules. The seller still serves two classes of customers with different delivery lead-time requirements: short and long lead-time customers. The orders from both classes of customer can be backlogged at certain costs. The replenishments are fixed and arrive one cycle late. Upon the arrival of each replenishment, the seller must determine the fulfillment of backlogged orders. Three priority rules are considered: (1) FCFS, (2) myopic short first-served (MSFS), and (3) strategic short first-served (SSFS). FCFS does not discriminate between orders from two classes of customers and clears outstanding orders purely based on their arrival times. In contrast, MSFS clears orders from short lead-time customers first, and SSFS goes one step further to reserve inventory for future short lead-time customers. We characterize the optimal inventory commitment policies under the three priority rules and further evaluate their performances with different initial-inventory and replenishment settings.

In the remainder of this chapter, Sect. 6.2 presents the formulation of the problem and three backlog-clearing priority rules. In Sect. 6.3, we derive the optimal inventory commitment policies for the three priority rules. Section 6.4 compares the performances of the three rules through numerical studies. Section 6.5 concludes the chapter.

6.2 Problem Formulation and Three Backlog-Clearing Priority Rules

The model setting mostly follows that seen in Chap. 3 with several notable changes. First, we assume that there exists a one-cycle replenishment lead time and that the replenishment decisions are given. That is, at the beginning of cycle i, the seller makes a replenishment of Q_i, which arrives at the beginning of cycle $i + 1$. Fixed replenishment decisions allow us to focus on comparing the three backlog-clearing priority rules. Second, in addition to long lead-time orders, we assume that short lead-time orders can be backlogged if the seller runs out of on-hand inventory and cannot satisfy them immediately. However, the seller incurs a penalty cost c_s (loss of goodwill) when backlogging a short lead-time order. Furthermore, at the beginning of the next cycle, the seller incurs a penalty cost c_{pl} for each backlogged long lead-time order if it cannot be filled and a possibly different penalty cost c_{ps} (explained in more detail later) for each unsatisfied short lead-time order. The third change is that the holding cost is now calculated in each period rather than at the end of a cycle. This means that the seller can save on the holding cost if it chooses to deliver a long lead-time order immediately.

At the beginning of a cycle, when the inventory is replenished, the seller chooses how to clear outstanding backlogged orders. Three backlog-clearing priority rules are considered: FCFS clears backlogged orders in the sequence of their arrival times; MSFS clears short ahead of long lead-time orders; and under SSFS, outstanding and future short lead-time orders have higher priority over outstanding long lead-time orders. In the remainder of this section, we detail the three priority rules. Throughout our analysis, we consider the cost-minimization problem.

6.2.1 The FCFS Model

We first write out the dynamic programming equations for the seller within a cycle. Let $V_i^F(t,n,m)$ be the cost-to-go function in period t for cycle i. Suppose that there are I cycles. For $1 \le i \le I-1$, we have

$$\begin{aligned} V_i^F(t,n,m) =& \pi_0[V_i^F(t+1,n,m)+hn]+\pi_s[V_i^F(t+1,n-1,m)+h(n-1)] \\ &+\pi_f \min_{u\in\{0,1\}}\{V_i^F(t+1,n-u,m+1-u)+h(n-u)\}, \text{ if } n\ge 1, \end{aligned} \tag{6.1}$$

$$\begin{aligned} V_i^F(t,n,m) =& \pi_0 V_i^F(t+1,n,m)+\pi_s[V_i^F(t+1,n-1,m)+c_s] \\ &+\pi_f V_i^F(t+1,n,m+1), \text{ if } n\le 0, \end{aligned} \tag{6.2}$$

where $u_i(t)$ still represents the commitment control imposed on the incoming long lead-time order. When the seller runs out of on-hand inventory, short lead-time orders are backlogged rather than lost; thus, n can take negative values, as shown in Eq. 6.2. The absolute value of a negative n represents the number of backlogged short lead-time orders.

We now look at the inventory dynamics between cycles. At the beginning of cycle $i+1$, the replenishment order placed in cycle i, denoted by Q_i, arrives. Under FCFS, outstanding backlogged orders are filled as much as possible from the leftover inventory from cycle i and the newly replenished inventory Q_i. If there is any shortage, i.e., when $n_i(T)+Q_i<m_i(T)$, the seller further backlogs these unfilled orders regardless of their lead-time preferences and incurs the same penalty $c_p=c_{ps}=c_{pl}$. The initial inventory position at the beginning of cycle $i+1$ is thus updated as $n_i(T)+Q_i-m_i(T)$.

$$V_i^F(T,n,m)=c_p(m-n-Q_i)^+ + V_{i+1}^F(1,n+Q_i-m,0), \tag{6.3}$$

where $c_p(m-n-Q_i)^+$ represents the penalty for further backlogging unfilled short and long lead-time orders when the seller does not have sufficient inventory. All unfilled orders are merged together and reflected in a negative inventory position. The amount of outstanding backlogged long lead-time orders is reset to zero.

For the last cycle I, because there is no future inventory replenishment, long lead-time orders are either filled or rejected, and short lead-time orders are filled to the extent possible. That is, the seller no longer commits to fulfilling orders in the "next" cycle, and $m=0$ throughout cycle I. For $1\le t\le T-1$,

$$V_I^F(t,n,0) = \pi_0[V_I^F(t+1,n,0)+hn] + \pi_s[V_I^F(t+1,n-1,0)+h(n-1)] \\ + \pi_f \min_{u\in\{0,1\}} \{V_i^F(t+1,n-u,0)+h(n-u)\}, \text{ if } n \geq 1, \quad (6.4)$$

$$V_I^F(t,n,0) = \pi_0 V_I^F(t+1,n,0) + \pi_s[V_I^F(t+1,n-1,0)+c_s] \\ + \pi_f V_i^F(t+1,n,0), \text{ if } n \leq 0. \quad (6.5)$$

And,

$$V_I^F(T,n,0) = S(n,0), \quad (6.6)$$

where $S(n,0)$ is the penalty/salvage function. We assume that the function satisfies the following properties.

Assumption 6.1 $S(n-1,m) - S(n,m+1)$ is non-increasing in n and independent of m; $S(n-1,m) - S(n,m)$ is non-increasing in n and non-decreasing in m; and $S(n,m) - S(n,m+1)$ is non-increasing in m.

Consider a linear example that $S(n,m) = s_s(-n)^+ + s_f m$, where s_s and s_f are penalty costs for not being able to meet customer orders. It is straightforward to check that $S(n,m)$ satisfies Assumption 6.1.

6.2.2 *The MSFS Model*

In the FCFS model, upon the arrival of an inventory replenishment, the seller fills the outstanding backlogged orders to the extent possible. Unfilled short and long lead-time orders are backlogged further and incur the same penalty cost c_p. In the next two models, MSFS and SSFS, we consider priority rules that give higher priority to outstanding short lead-time orders in the sense that unfilled short lead-time orders incur a higher penalty cost than their long lead-time counterparts, i.e., $c_{ps} > c_{pl}$. We first introduce the MSFS model. In the MSFS model, the seller clears backlogged long lead-time orders only after clearing all of the outstanding short lead-time orders. (Note that when $c_{ps} = c_{pl}$, the MSFS model is equivalent to the FCFS model regardless of the higher priority given to backlogged short lead-time orders.)

Let $V_i^M(t,n,m)$ be the cost-to-go function in period t for cycle i. Within a cycle, the dynamic programming recursion is the same as that in the FCFS model.

$$V_i^M(t,n,m) = \pi_0[V_i^M(t+1,n,m)+hn] + \pi_s[V_i^M(t+1,n-1,m)+h(n-1)] \\ + \pi_f \min_{u\in\{0,1\}} \{V_i^M(t+1,n-u,m+1-u)+h(n-u)\}, \text{ if } n \geq 1, \quad (6.7)$$

$$V_i^M(t,n,m) = \pi_0 V_i^M(t+1,n,m) + \pi_s(V_i^M(t+1,n-1,m) + c_s) \\ +\pi_f V_i^M(t+1,n,m+1), \text{ if } n \le 0. \tag{6.8}$$

In cycle i, the ending inventory to be carried over to the next cycle is determined by $n_i(T)$, which can be either positive or negative, $m_i(T)$, and Q_i. There are three possible scenarios: (i) $n_i(T) + Q_i \le 0$; (ii) $0 \le n_i(T) + Q_i \le m_i(T)$; and (iii) $n_i(T) + Q_i \ge m_i(T)$. In scenario (i), the seller cannot satisfy all of the backlogged short lead-time orders; the ending inventory position is thus $n_i(T) + Q_i$, and the outstanding long lead-time orders remain unchanged. The remaining short and long lead-time orders incur unit penalty costs c_{ps} and c_{pl}, respectively. In scenario (ii), the seller can satisfy all of the backlogged short lead-time orders, but only part of the outstanding long lead-time orders. The ending inventory position is 0, and the remaining long lead-time orders equal $m_i(T) - n_i(T) - Q_i$. In scenario (iii), the seller can fill all of the backlogged short and long lead-time orders. For $1 \le i \le I-1$,

$$V_i^M(T,n,m) = \begin{cases} c_{ps}(-n-Q_i) + c_{pl}m + V_{i+1}^M(1,n+Q_i,m), & \text{if } n+Q_i \le 0; \\ c_{pl}(m-n-Q_i) + V_{i+1}^M(1,0,m-n-Q_i), & \text{else if } 0 \le n+Q_i \le m; \\ V_{i+1}^M(1,n+Q_i-m,0), & \text{otherwise, } m \le n+Q_i. \end{cases} \tag{6.9}$$

In the last cycle, long lead-time orders are either filled or rejected, and short lead-time orders are filled to the extent possible.

$$V_I^M(t,n,m) = \pi_0[V_I^M(t+1,n,m) + hn] + \pi_s[V_I^M(t+1,n-1,m) + h(n-1)] \\ +\pi_f \min_{u\in\{0,1\}} \{V_i^M(t+1,n-u,m) + h(n-u)\}, \text{ if } n \ge 1 \tag{6.10}$$

$$V_I^M(t,n,m) = \pi_0 V_I^M(t+1,n,m) + \pi_s[V_I^M(t+1,n-1,m) + c_s] \\ +\pi_f V_i^M(t+1,n,m), \text{ if } n \le 0. \tag{6.11}$$

The terminal condition for the last cycle I is the same as (6.6) with the superscript F replaced by M. That is,

$$V_I^M(T,n,m) = S(n,m), \tag{6.12}$$

where $S(n,m)$ satisfies Assumption 6.1.

6.2.3 The SSFS Model

In the MSFS model, after all of the outstanding short lead-time orders are satisfied, the remaining inventory is used to fill as many outstanding long lead-time orders as possible. The leftover inventory is then carried over to the next cycle. If the inventory is not enough to clear all of the backlogged orders, the seller starts with a negative inventory position in the next cycle. As a result, all of the short and long lead-time orders in the next cycle have to be backlogged. We thus consider a priority rule under which the seller may choose not to clear all of the outstanding long lead-time orders, but to backlog some of them further. The seller thus can keep some strategic inventory for the next cycle at the expense of delaying long lead-time orders. We call this model the strategic short first-served (SSFS) model.

Let $V_i^S(t,n,m)$ be the cost-to-go function in period t for cycle i. Within a cycle, the dynamic programming recursion is the same as those in the FCFS and the MSFS models.

$$\begin{aligned} V_i^S(t,n,m) = &\pi_0[V_i^S(t+1,n,m)+hn]+\pi_s[V_i^S(t+1,n-1,m)+h(n-1)] \\ &+\pi_f \min_{u\in\{0,1\}}\{V_i^S(t+1,n-u,m+1-u)+h(n-u)\}, \text{ if } n\geq 1, \end{aligned} \tag{6.13}$$

$$\begin{aligned} V_i^S(t,n,m) = &\pi_0 V_i^S(t+1,n,m)+\pi_s[V_i^S(t+1,n-1,m)+c_s] \\ &+\pi_f V_i^S(t+1,n,m+1), \text{ if } n\leq 0. \end{aligned} \tag{6.14}$$

At the beginning of cycle $i+1, i=1,\cdots,I-1$, upon the arrival of the replenishment of Q_i, the seller allocates only a part of the available on-hand inventory, denoted by y, to fill the outstanding long lead-time orders. Note that y is constrained by the on-hand inventory level and the number of outstanding long lead-time orders, i.e., $0\leq y\leq \min\{m_i(T),(n_i(T)+Q_i)^+\}$. The value of y can be obtained by solving the following optimization problem. For $1\leq i\leq I-1$,

$$\begin{aligned} &V_i^S(T,n,m) \\ &= c_{ps}(-n-Q_i)^+ + \min_{0\leq y\leq \min\{m,(n+Q_i)^+\}}\{c_{pl}(m-y)+V_{i+1}^S(1,n+Q_i-y,m-y)\} \\ &= \begin{cases} c_{ps}(-n-Q_i)+c_{pl}m+V_{i+1}^S(1,n+Q_i,m), & \text{if } n+Q_i\leq 0; \\ \min\limits_{0\leq y\leq n+Q_i}\{c_{pl}(m-y)+V_{i+1}^S(1,n+Q_i-y,m-y)\}, & \text{else if } 0\leq n+Q_i\leq m; \\ \min\limits_{0\leq y\leq m}\{c_{pl}(m-y)+V_{i+1}^S(1,n+Q_i-y,m-y)\}, & \text{otherwise, } m\leq n+Q_i, \end{cases} \end{aligned} \tag{6.15}$$

where $c_{ps}(-n-Q_i)^+$ and $c_{pl}(m-y)$ represent the penalty costs for outstanding short and long lead-time orders, respectively.

In the last cycle,

$$\begin{aligned} V_I^S(t,n,m) = &\pi_0[V_I^S(t+1,n,m)+hn]+\pi_s[V_I^S(t+1,n-1,m)+h(n-1)] \\ &+\pi_f \min_{u\in\{0,1\}} V_i^S(t+1,n-u,m)+h(n-u), \text{ if } n\geq 1, \end{aligned} \tag{6.16}$$

$$\begin{aligned} V_I^S(t,n,m) = &\pi_0 V_I^S(t+1,n,m)+\pi_s[V_I^S(t+1,n-1,m)+c_s] \\ &+\pi_f V_i^S(t+1,n,m), \text{ if } n\leq 0. \end{aligned} \tag{6.17}$$

The terminal condition is

$$V_I^S(T, n, m) = S(n, m), \tag{6.18}$$

where $S(n, m)$ satisfies Assumption 6.1.

6.3 Optimal Inventory Commitment Policies

In all three of the models introduced previously, the seller imposes commitment control over incoming long lead-time orders. One major difference between these models and those discussed in the previous chapters is that short lead-time orders can also be backlogged now. This, from a cost-minimization perspective, should induce the seller to be less willing to choose to deliver a long lead-time order immediately because backlogging a short lead-time order is more costly. Nevertheless, in our analysis in this section, we show that the structural properties and optimal policies still remain very much the same as those we derive for the basic model in Chap. 3.

6.3.1 The FCFS Model

From (6.1), we know that the commitment-control policy depends on the values of $V_i^F(t+1, n-1, m)$ and $V_i^F(t+1, n, m+1)$. The former corresponds to the action that commits a unit of on-hand inventory to the long lead-time order, and the latter corresponds to the action that backlogs the order to the next cycle. We first introduce the following definition.

Definition 6.1 The marginal cost of inventory commitment is defined as the cost difference between committing a unit of on-hand inventory and backlogging the order to the next cycle, i.e., $V_i^F(t, n-1, m) - V_i^F(t, n, m+1)$.

We can obtain the following lemma. All proofs are relegated to the appendix.

Lemma 6.1 *In period t of cycle i $(1 \le i \le I-1)$, the marginal cost of inventory commitment is (a) non-increasing in n when $n \ge 1$ and 0 when $n \le 0$, (b) independent of m, and (c) independent of Q_i.*

Part (a) of the lemma indicates that when the seller still has some on-hand inventory, the higher the on-hand inventory level is, the better it is to commit a unit of on-hand inventory to a long lead-time order. Furthermore, part (b) reveals that this decision is independent of the outstanding long lead-time orders. When making the commitment decision, the seller takes into account the trade-off between the holding cost and the possible penalty for failing to fill a short lead-time order. As there is no penalty cost for backlogging long lead-time orders, m does not play any role in the trade-off. Furthermore, regardless of whether the seller fills the order immediately or in the next cycle, the difference between the inventory position and outstanding orders, i.e., $n - m$, does not change, which means that the penalty cost for running out of replenished inventory (the first term in (6.3)) does not change. Hence, Q_i also does not come into play in the trade-off, and part (c) holds. To characterize the optimal commitment policy, we define the commitment level as follows.

Definition 6.2 For each $m + 1$, the commitment level, denoted by $C_L(i, t, m + 1)$, is 0 if $V_i^F(t + 1, 0, m) - V_i^F(t + 1, 1, m + 1) \leq h$; otherwise, the commitment level is obtained as the maximum value of n for which $V_i^F(t + 1, n - 1, m) - V_i^F(t + 1, n, m + 1) \geq h$.

Based on this definition and Lemma 6.1, we can characterize the optimal commitment policy for long lead-time orders.

Theorem 6.1 *Under FCFS, in period t of cycle i, the following holds.*

(a) The optimal commitment policy is characterized by a commitment level $C_L(i, t, m)$.

(b) The commitment level $C_L(i, t, m)$ *is independent of m and* Q_i.

(c) The optimal commitment policy is to commit one unit of on-hand inventory to a long lead-time order if $n(t) > C_L(i, t, m)$ *and to backlog the order to the next cycle otherwise.*

The following theorem provides some further characterization of the optimal commitment level.

Theorem 6.2 $C_L(i, t, m) + 1 \geq C_L(i, t - 1, m) \geq C_L(i, t, m)$.

Theorem 6.2 shows how the optimal commitment policy changes with time. It also facilitates the calculation of the commitment level. The intuition for the results can be explained as follows. As t increases, there is less time remaining and fewer potential short lead-time orders. The seller is thus more willing to commit one unit of on-hand inventory to a long lead-time order. As for the upper bound given by the first inequality, it can be considered as a direct result of the unitary demand arrival setting.

In addition to Lemma 6.1, we can derive the following properties of the cost-to-go function.

Lemma 6.2 *In period t of cycle i,* $V_i^F(t, n, m)$ *is (a) convex in n, (b) convex in m, and (c) submodular in n and m.*

The convexity property in (a) implies that the greater the inventory position, the smaller the marginal cost reduction. The convexity property in (b) implies that the more outstanding long lead-time orders there are, the smaller the marginal cost reduction. The submodularity property in (c) implies that the inventory position and outstanding orders are economic complements. Based on Lemmas 6.1 and 6.2, it is straightforward to extend preceding analysis to cycle $i - 1$.

6.3.2 The MSFS Model

In the MSFS model, the seller clears short lead-time orders ahead of long lead-time orders upon the arrival of an inventory replenishment. Still using the definitions introduced in the discussion of the FCFS model, we can prove the following result.

Lemma 6.3 *In each cycle i, the following holds.*

(a) The marginal cost of inventory commitment, i.e., $V_i^M(T, n-1, m) - V_i^M(T, n, m+1)$, is non-increasing in n when $n + Q_i \le 0$, independent of n when $n + Q_i \ge 0$, and independent of m.

(b) $V_i^M(T, n-1, m) - V_i^M(T, n, m)$ is non-increasing in n and non-decreasing in m.

(c) $V_i^M(T, n, m) - V_i^M(T, n, m+1)$ is non-increasing in m.

(d) $V_i^M(T, n-1, m) - V_i^M(T, n, m+1)$ is non-increasing in Q_i.

Recall that within a cycle, the inventory dynamics in the MSFS model remain the same as those in the FCFS model. We can thus follow the same approaches used in the proofs of Lemmas 6.1 and 6.2 to derive the structural properties of $V_i^M(t, n, m)$. However, note that the marginal cost of inventory commitment now depends on Q_i and is not independent of Q_i as in the FCFS model. MSFS clears short lead-time orders first and changing Q_i can lead to different penalty costs in (6.9) at the end of each cycle.

With the properties described in Lemma 6.3, we can characterize the optimal commitment policy as follows.

Theorem 6.3 *Under MSFS, (a) the optimal commitment policy is characterized by a commitment level $C_L(i, t, m)$; (b) the commitment level $C_L(i, t, m)$ is independent of m and non-decreasing in Q_i; and (c) the optimal commitment policy is to commit one unit of on-hand inventory to the long lead-time order immediately if $n(t) > C_L(i, t, m)$ and to backlog it to the next cycle otherwise.*

Theorem 6.3 reveals that the optimal commitment policy under MSFS is of a threshold type. The commitment level is non-decreasing in Q_i. Under MSFS, a larger replenishment order reduces the possibility of running out of on-hand inventory and thereby reduces potential penalty costs for further backlogging of unfilled orders.

6.3.3 The SSFS Model

The dynamic programming recursion within a cycle in the SSFS model is the same as that in the FCFS model. We thus focus on the dynamic programming equations between cycles. We prove the following properties by backward induction: (a) the marginal cost of inventory commitment, $V_{i+1}^S(t, n-1, m) - V_{i+1}^S(t, n, m+1)$, is non-increasing in n, independent of m, and non-increasing in Q_i; (b) $V_{i+1}^S(t, n-1, m) - V_{i+1}^S(t, n, m)$ is non-increasing in n and non-decreasing in m; and (c)

$V_{i+1}^S(t,n,m) - V_{i+1}^S(t,n,m+1)$ is non-increasing in m. For the last cycle, it is obvious that the terminal function (6.18) satisfies the preceding properties. We now assume that the foregoing properties hold for $V_{i+1}^S(T,n,m)$. Following a similar approach to that in the proofs of Lemmas 6.1 and 6.2, we can prove that $V_{i+1}^S(t,n,m)$ $(1 \le t \le T-1)$ preserves these properties. To conclude the backward induction, in what follows, we prove that these properties can also be carried over to $V_i^S(T,n,m)$.

We first derive the seller's optimal backlog-clearing policy upon the arrival of Q_i. Let

$$F(y,m) = c_{pl}(m-y) + V_{i+1}^S(1, n+Q_i-y, m-y), \tag{6.19}$$

where y represents the inventory that is used to clear backlogged long lead-time orders and $0 \le y \le \min\{m, (n+Q_i)^+\}$. Let $n_{i+1}(1) = n + Q_i - y$, so $y = n + Q_i - n_{i+1}(1)$. Denote $n_{i+1}^*(1)$ as the minimizer of (6.19). That is,

$$n_{i+1}^*(1) = \arg\min_{n_{i+1}(1)} \{c_{pl}[m-n-Q_i+n_{i+1}(1)] + V_{i+1}^S(1, n_{i+1}(1), m+n_{i+1}(1)-Q_i-n)\}. \tag{6.20}$$

The following lemma defines the optimal y, denoted as y^*.

Lemma 6.4 *$F(y,m)$ is convex in y and $n_{i+1}^*(1)$ is independent of n and m. When $n_{i+1}^*(1) \ge 0$,*

$$y^* = \begin{cases} 0, & \text{if } n+Q_i \le 0 \text{ or } 0 \le n+Q_i \le n_{i+1}^*(1); \\ n+Q_i-n_{i+1}^*(1), & \text{if } n+Q_i \ge 0 \text{ and } n+Q_i-\min\{m, n+Q_i\} \le n_{i+1}^*(1) \le n+Q_i; \\ m, & \text{if } n+Q_i \ge 0 \text{ and } n_{i+1}^*(1) \le n+Q_i-\min\{m, n+Q_i\}. \end{cases}$$

Otherwise, $y^* = \min\{m, (n+Q_i)^+\}$.

Note that $n_{i+1}^*(1)$ represents the optimal inventory reserved for future short lead-time orders and is independent of n and m. If $n_{i+1}^*(1) < 0$, it means that the seller should not reserve any inventory. This could happen, e.g., if Q_i is large or the inventory holding cost is high. In this case, the SSFS model is equivalent to the MSFS model.

After we characterize the optimal inventory policy between cycles, we can prove that the structural properties we derive for the cost-to-go function in cycle $i+1$ can be carried over to cycle i.

Lemma 6.5 *In cycle i, the following holds.*

(a) $V_i^S(T,n-1,m) - V_i^S(T,n,m+1)$ is non-increasing in n, independent of m, and non-increasing in Q_i.

(b) $V_i^S(T,n-1,m) - V_i^S(T,n,m)$ is non-increasing in n and non-decreasing in m.

(c) $V_i^S(T,n,m) - V_i^S(T,n,m+1)$ is non-increasing in m.

Note that the marginal cost of inventory commitment, $V_i^S(T,n-1,m) - V_i^S(T,n,m+1)$, is independent of n when $n+Q_i \ge 0$ in the MSFS model as stated in Lemma 6.3. In contrast, in the SSFS model, when $n+Q_i \ge 0$, $V_i^S(T,n-1,m) -$

$V_i^S(T, n, m+1)$ is non-increasing in n. The seller may further backlog some outstanding orders to reserve on-hand inventory for future demands. Hence, a greater on-hand inventory level can reduce unfilled outstanding orders carried over to the next cycle, and the difference $V_i^S(T, n-1, m) - V_i^S(T, n, m+1)$ is smaller.

We are now ready to characterize the inventory commitment policy.

Theorem 6.4 *Under SSFS, the optimal commitment policy is characterized by a commitment level $C_L(i, t, m)$, which is independent of m and non-decreasing in Q_i. The optimal commitment policy is to allocate on-hand inventory to a long lead-time order if $n(t) > C_L(i, t, m)$ and to backlog it to the next cycle otherwise.*

6.4 Comparative Studies

In this section, we compare the optimal commitment policies under MSFS and SSFS and identify the conditions that affect their performances. The results elucidate the conditions under which a priority rule makes more sense. We first study the optimal commitment policies under MSFS and SSFS, and then further compare their performances.

To compare the commitment policies under MSFS and SSFS, we look at the marginal cost of inventory commitment as defined in Definition 6.1. Noting that the marginal cost of inventory commitment is independent of m in both the MSFS and SSFS models, the following result holds.

Lemma 6.6 *For given n, $V_i^S(t, n-1, m) - V_i^S(t, n, m+1) \geq V_i^M(t, n-1, m) - V_i^M(t, n, m+1)$.*

Given that the marginal cost of inventory commitment determines the commitment level, the following theorem is straightforward.

Theorem 6.5 *In each period, the commitment level under SSFS is no less than that under MSFS.*

Theorem 6.5 indicates that given the same level of on-hand inventory, the seller is more likely to backlog long lead-time orders to the next cycle under SSFS.

We now study the system performance with varying initial inventory and replenishment quantities under the two priority rules using a two-cycle model. In the SSFS model, at the beginning of the second cycle, the seller reserves inventory for the demands in this cycle, so there is an inventory rationing decision. The commitment policies and the cost-to-go functions in the two models can be calculated using backward induction. In both models, we assume that $S(n, m) = s_s(-n)^+ + s_f m$, where $s_f = c_{pl} + c$ and $s_s = \beta * s_f$. We list the parameter values in Table 6.1.

We draw the contour curves of the cost-to-go functions, i.e., $V_1^M(1, n, 0)$ and $V_1^S(1, n, 0)$, in relation to the initial inventory and the replenishment in Figs. 6.1 and 6.2, respectively. Figure 6.3 further quantifies the difference between $V_1^M(1, n, 0)$

Table 6.1 Parameter Values

c_s	c_{pl}	c_{ps}	h	c	β	π_s	π_f	π_0
2	1	2	0.1	5	1.5	0.35	0.35	0.3

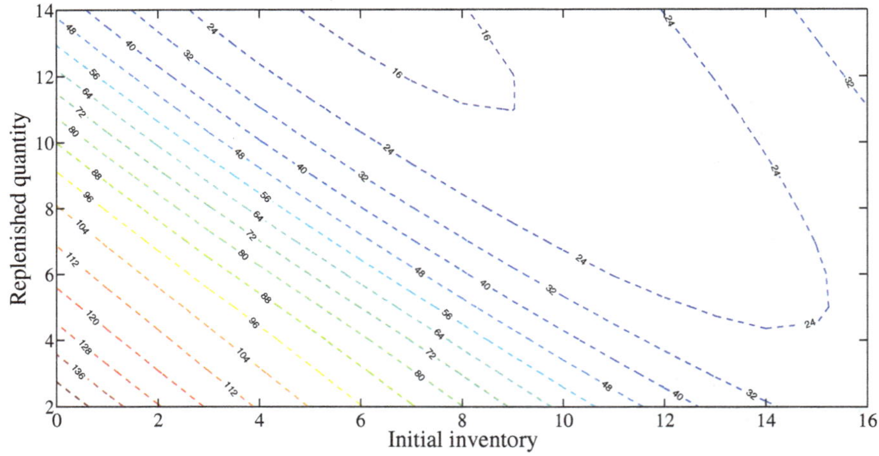

Fig. 6.1 Contour curves of operating cost under MSFS

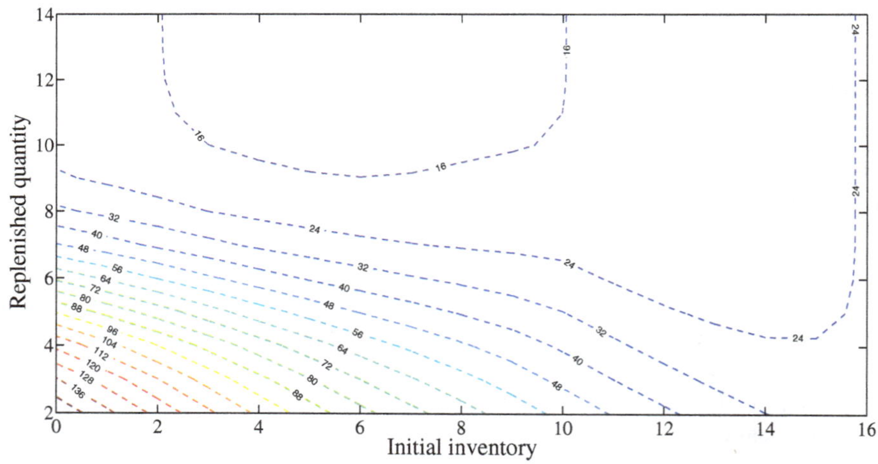

Fig. 6.2 Contour curves of operating cost under SSFS

and $V_1^S(1, n, 0)$ as $(V_1^M(1, n, 0) - V_1^S(1, n, 0))/V_1^S(1, n, 0)$ (i.e., the relative performance improvement of SSFS). From Fig. 6.3, it can be observed that the improvement is significant when the initial inventory is low and the replenishment is large. This observation agrees with intuition because the strategic inventory reservation by SSFS makes more of a difference in this case.

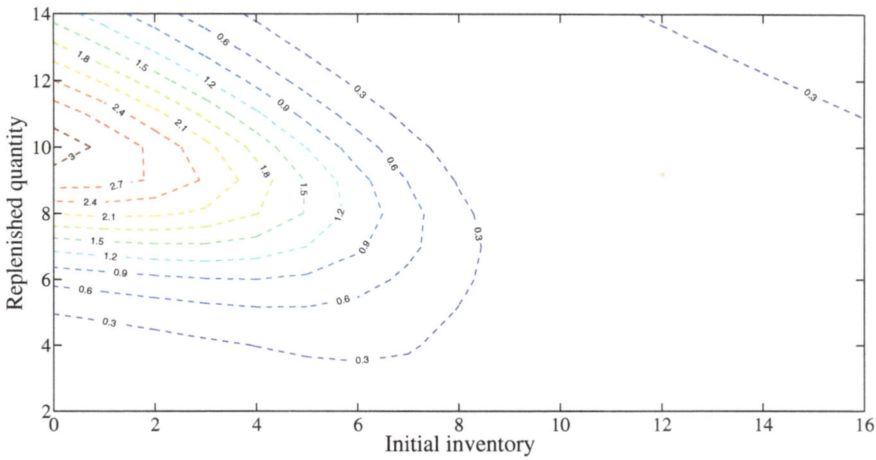

Fig. 6.3 Contour curves of the cost difference under MSFS and SSFS

To investigate whether the order retention is worthwhile, we include the lost-sales model as studied in Chap. 3 in the comparison. The setting is as follows. Within a cycle, the dynamic programming recursion for the lost-sales model is the same as that for the SSFS model. However, between the cycles, in the lost-sales model, backlogged short lead-time orders (i.e., those short lead-time orders that arrive during stock-out) are lost, incurring a penalty cost that equals $-s_s n$. Outstanding long lead-time orders are filled by the on-hand inventory and replenishment. Unfilled long lead-time orders incur a penalty cost that equals $s_f(m - Q)^+$ when $n \leq 0$ and $s_f(m - Q - n)^+$ when n is positive. Hence, the seller carries over only on-hand inventory between cycles in the lost-sales model. Let $V_i^F(t, n, m)$ be the cost-to-go function of the lost-sales model. Thus, for $1 \leq i \leq I - 1$,

$$V_i^L(T, n, m)$$
$$= \begin{cases} -s_s n + s_f(m - Q)^+ + V_{i+1}^L(1, (Q - m)^+, (m - Q)^+), & \text{if } n \leq 0; \\ s_f(m - Q - n)^+ + V_{i+1}^L(1, (n + Q - m)^+, (m - Q - n)^+), & \text{otherwise.} \end{cases}$$

Figure 6.4 shows the comparison between the lost-sales model and the SSFS model using the value of $(V_1^F(1, n, 0) - V_1^S(1, n, 0))/V_1^S(1, n, 0)$. Under SSFS, the seller pays c_{ps} (c_{pl}) to retain unfilled short (long) lead-time orders. The first observation is that the lost-sales model does not necessarily underperform. This is most likely to be true when the replenishment quantity is low. In this case, simply backlogging unfilled orders further does not improve the overall performance. The second observation is that SSFS outperforms the lost-sales model when the initial inventory is low and the replenishment quantity is high. In this case, further backlogging in the SSFS model helps to reduce the penalty cost, which is inevitable in the lost-sales case. This observation is also consistent with the real-world practices adopted by car

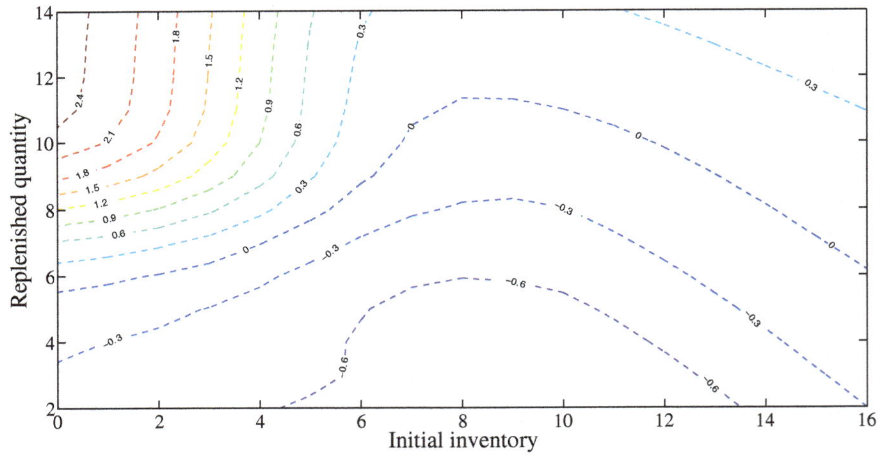

Fig. 6.4 Contour curves of the cost difference under Lost-sales and SSFS

dealerships. These dealerships are most willing to pay incentives to retain unfilled orders when the on-hand inventory is low but a large amount of replenishment is on its way (Liang et al. 2013).

6.5 Concluding Remarks

In this chapter, we focus on the priority rules in clearing backlogged orders. Three rules are considered: FCFS treats all backlogged orders uniformly; MSFS clears short ahead of long lead-time orders; and SSFS allows the seller to reserve inventory for future short lead-time orders. We characterize the optimal inventory commitment policies under all three priority rules. We also compare the performances under MSFS, SSFS, and the lost-sales model. We find that the relative performance of SSFS improves when the initial inventory is low and the replenishment quantity is large.

Appendix

The proofs of the results in the FCFS and MSFS models are similar to those in Chap. 3 and are therefore omitted.

Proof of Lemma 6.4

Proof According to the assumption in the first paragraph of this subsection, i.e., that $V_{i+1}^S(T,n,m)$ satisfies property (a), we know that $V_{i+1}^S(T,n-1,m) - V_{i+1}^S(T,n,m+1)$ is non-increasing in n and independent of m. We can prove that $F(y,m)$ is convex in y based on (6.19):

$$\begin{aligned}
&F(y,m) - F(y-1,m)\\
&= -c_{pl} + V_{i+1}^S(T,n+Q_i-y,m-y) - V_{i+1}^S(T,n+Q_i-y+1,m-y+1)\\
&\le -c_{pl} + V_{i+1}^S(T,n+Q_i-y-1,m-y) - V_{i+1}^S(T,n+Q_i-y,m-y+1)\\
&= -c_{pl} + V_{i+1}^S(T,n+Q_i-y-1,m-y-1) - V_{i+1}^S(T,n+Q_i-y,m-y)\\
&= F(y+1,m) - F(y,m),
\end{aligned}$$

where the inequality results because $V_{i+1}^S(T,n-1,m) - V_{i+1}^S(T,n,m+1)$ is non-increasing in n and the second equality holds because $V_{i+1}^S(T,n-1,m) - V_{i+1}^S(T,n,m+1)$ is independent of m. It is clear that there exists an optimal y, denoted as y^*. Furthermore, as obtained at the beginning of this subsection, because $V_{i+1}^S(t,n-1,m) - V_{i+1}^S(t,n,m+1)$ is independent of m, we know that $V_{i+1}^S(1,n_{i+1}(1),m+n_{i+1}(1)-Q_i-n) - V_{i+1}^S(1,n_{i+1}(1)-1,m+n_{i+1}(1)-1-Q_i-n)$ is independent of m and n. Hence, $n_{i+1}^*(1)$ as defined in (6.20) is independent of n and m.

Proof of Lemma 6.5

Proof When $n+Q_i \le 0$, (6.15) is reduced to

$$V_i^S(T,n,m) = c_{ps}(-n-Q_i) + c_{pl}m + V_{i+1}^S(1,n+Q_i,m).$$

In this case, $n-1+Q_i \le 0$. It is easy to show that $V_i^S(T,n,m)$ satisfies properties (a), (b), (c), and (d). Otherwise, when $n+Q_i \ge 1$ and $n-1+Q_i \ge 0$, based on (6.15), we have

$$\begin{aligned}
V_i^S(T,n-1,m) &= \min_{0\le y\le \min\{m,n-1+Q_i\}} \{c_{pl}(m-y) + V_{i+1}^S(1,n-1+Q_i-y,m-y)\}\\
&= \min_{1\le y\le \min\{m+1,n+Q_i\}} \{c_{pl}(m-y+1) + V_{i+1}^S(1,n+Q_i-y,m+1-y)\}\\
&= \min_{1\le y\le \min\{m+1,n+Q_i\}} F(y,m+1),
\end{aligned} \tag{6.21}$$

$$\begin{aligned}
V_i^S(T,n,m+1) &= \min_{0\le y\le \min\{m+1,n+Q_i\}} \{c_{pl}(m+1-y) + V_{i+1}^S(1,n+Q_i-y,m+1-y)\}\\
&= \min_{0\le y\le \min\{m+1,n+Q_i\}} F(y,m+1),
\end{aligned} \tag{6.22}$$

$$\begin{aligned} V_i^S(T, n-1, m+1) &= \min_{1 \le y \le \min\{m+2, n+Q_i\}} \{c_{pl}(m-y+2) + V_{i-1}^S(1, n+Q_i-y, m+2-y)\} \\ &= \min_{1 \le y \le \min\{m+2, n+Q_i\}} F(y, m+2), \end{aligned} \tag{6.23}$$

and

$$\begin{aligned} V_i^S(T, n, m+2) &= \min_{0 \le y \le \min\{m+2, n+Q_i\}} \{c_{pl}(m+2-y) + V_{i+1}^S(1, n+Q_i-y, m+2-y)\} \\ &= \min_{0 \le y \le \min\{m+2, n+Q_i\}} F(y, m+2). \end{aligned} \tag{6.24}$$

From Lemma 6.4, we know that $F(y, m)$ is convex in y and the minimizer is independent of m, i.e., $F(y, m) - F(y-1, m)$ is independent of m. Let α^* be the minimizer of $F(y, m)$. We now evaluate (6.21)–(6.24) according to the value of α^*. When $\alpha^* \le 0$,

$$\begin{aligned} & V_i^S(T, n-1, m) - V_i^S(T, n, m+1) \\ &= F(1, m+1) - F(0, m+1) \\ &= F(1, m+2) - F(0, m+2) \\ &= V_i^S(T, n-1, m+1) - V_i^S(T, n, m+2), \end{aligned} \tag{6.25}$$

where the second inequality is due to the fact that $F(y, m) - F(y-1, m)$ is independent of m. Similarly, when $1 \le \alpha^* \le \min\{m+1, n+Q_i\}$,

$$\begin{aligned} & V_i^S(T, n-1, m) - V_i^S(T, n, m+1) \\ &= F(\alpha^*, m+1) - F(\alpha^*, m+1) \\ &= F(\alpha^*, m+2) - F(\alpha^*, m+2) \\ &= V_i^S(T, n-1, m+1) - V_i^S(T, n, m+2). \end{aligned} \tag{6.26}$$

When $\alpha^* \ge \min\{m+2, n+1+Q_i\}$,

$$\begin{aligned} & V_i^S(T, n-1, m) - V_i^S(T, n, m+1) \\ &= F(\min\{m+1, n+Q_i\}, m+1) - F(\min\{m+1, n+Q_i\}, m+1) \\ &= F(\min\{m+2, n+Q_i\}, m+2) - F(\min\{m+2, n+Q_i\}, m+2) \\ &= V_i^S(T, n-1, m+1) - V_i^S(T, n, m+2). \end{aligned} \tag{6.27}$$

The preceding equations show that $V_i^S(T, n-1, m) - V_i^S(T, n, m+1)$ is independent of m. Based on the definition of $F(y)$ in the case where $n + Q_i \ge 1$, we also know that

$$F(1, m+1) - F(0, m+1) = -c_{pl} + V_{i+1}^S(1, n+Q_i-1, m-1) - V_{i+1}^S(1, n+Q_i, m) \tag{6.28}$$

is non-increasing in Q_i. Hence, by (6.25)–(6.28), $V_i^S(T, n-1, m) - V_i^S(T, n, m+1)$ is non-increasing in Q_i.

Furthermore, based on (6.15),

$$\begin{aligned} V_i^S(T, n, m) &= \min_{0 \le y \le \min\{m, n+Q_i\}} \{c_{pl}(m-y) + V_{i+1}^S(1, n+Q_i-y, m-y)\} \\ &= \min_{0 \le y \le \min\{m, n+Q_i\}} F(y, m), \end{aligned} \tag{6.29}$$

and

$$\begin{aligned} V_i^S(T, n+1, m+1) &= \min_{0 \le y \le \min\{m+1, n+1+Q_i\}} \{c_{pl}(m+1-y) + V_{i+1}^S(1, n+1+Q_i-y, m+1-y)\} \\ &= \min_{-1 \le y \le \min\{m, n+Q_i\}} \{c_{pl}(m-y) + V_{i+1}^S(1, n+Q_i-y, m-y)\} \\ &= \min_{-1 \le y \le \min\{m, n+Q_i\}} F(y, m). \end{aligned} \tag{6.30}$$

Based on (6.21), (6.22), (6.29), and (6.30), we discuss the following cases. When $\alpha^* \ge 1$,

$$\begin{aligned} & V_i^S(T, n-1, m) - V_i^S(T, n, m+1) \\ &= F(\min\{\alpha^*, m+1, n+Q_i\}, m+1) - F(\min\{\alpha^*, m+1, n+Q_i\}, m+1)] \\ &= F(\min\{\alpha^*, m, n+Q_i\}, m) - F(\min\{\alpha^*, m, n+Q_i\}, m)] \\ &= V_i^S(T, n, m) - V_i^S(T, n+1, m+1). \end{aligned} \tag{6.31}$$

When $\alpha^* = 0$,

$$\begin{aligned} & V_i^S(T, n-1, m) - V_i^S(T, n, m+1) \\ &= F(1, m+1) - F(0, m+1) \\ &\ge F(0, m) - F(0, m) \\ &= V_i^S(T, n, m) - V_i^S(T, n+1, m+1). \end{aligned} \tag{6.32}$$

When $\alpha \le -1$,

$$\begin{aligned} & V_i^S(T, n-1, m) - V_i^S(T, n, m+1) \\ &= F(1, m+1) - F(0, m+1) \\ &\ge F(0, m) - F(-1, m) \\ &= V_i^S(T, n, m) - V_i^S(T, n+1, m+1), \end{aligned} \tag{6.33}$$

where the inequality is due to the convexity of $F(y, m)$ in y and the fact that $F(y, m) - F(y-1, m)$ is independent of m. Hence, based on (6.31)–(6.33), we know that $V_i^S(T, n-1, m) - V_i^S(T, n, m+1)$ is non-increasing in n.

Using the same approach, i.e., by checking different cases according to the value of α^*, we can prove the properties in (b) and (c). This completes the proof.

Proof of Lemma 6.6

Proof It is clear that $V_I^S(T, n-1, m) - V_I^S(T, n, m+1) \geq V_I^M(T, n-1, m) - V_I^M(T, n, m+1)$. Now, we assume that $V_i^S(T, n-1, m) - V_i^S(T, n, m+1) \geq V_i^M(T, n-1, m) - V_i^M(T, n, m+1)$ and prove that $V_{i-1}^S(T, n-1, m) - V_{i-1}^S(T, n, m+1) \geq V_{i-1}^M(T, n-1, m) - V_{i-1}^M(T, n, m+1)$.

First, we assume that $V_i^S(t+1, n-1, m) - V_i^S(t+1, n, m+1) \geq V_i^M(t+1, n-1, m) - V_i^M(t+1, n, m+1)$ and prove that $V_i^S(t, n-1, m) - V_i^S(t, n, m+1) \geq V_i^M(t, n-1, m) - V_i^M(t, n, m+1)$. From (6.8) and (6.11), it is clear that

$$V_i^S(t, n-1, m) - V_i^S(t, n, m+1) \geq V_i^M(t, n-1, m) - V_i^M(t, n, m+1), \text{ if } n \leq 0.$$

When $n = 1$, we must determine only whether the following inequality holds:

$$\begin{aligned} &V_{i+1}^S(t+1, n-1, m+1) + h(n-1) \\ &- \min\{V_{i+1}^S(t+1, n-1, m+1) + h(n-1), V_{i+1}^S(t+1, n, m+2) + hn\} \\ \geq\ &V_{i+1}^M(t+1, n-1, m+1) + h(n-1) \\ &- \min\{V_{i+1}^M(t+1, n-1, m+1) + h(n-1), V_{i+1}^M(t+1, n, m+2) + hn\}. \end{aligned}$$

We prove it by checking all possible cases. We have

$$\begin{aligned} &V_{i+1}^S(t+1, n-1, m+1) + h(n-1) \\ &- \min\{V_{i+1}^S(t+1, n-1, m+1) + h(n-1), V_{i+1}^S(t+1, n, m+2) + hn\} \\ \geq\ &V_{i+1}^S(t+1, n-1, m+1) + h(n-1) \\ &-\{V_{i+1}^S(t+1, n-1, m+1) + h(n-1)\} \\ \geq\ &V_{i+1}^M(t+1, n-1, m+1) + h(n-1) \\ &-\{V_{i+1}^M(t+1, n-1, m+1) + h(n-1)\}, \end{aligned}$$

and

$$\begin{aligned} &V_{i+1}^S(t+1, n-1, m+1) + h(n-1) \\ &- \min\{V_{i+1}^S(t+1, n-1, m+1) + h(n-1), V_{i+1}^S(t+1, n, m+2) + hn\} \\ \geq\ &V_{i+1}^S(t+1, n-1, m+1) + h(n-1) \\ &-\{V_{i+1}^S(t+1, n, m+2) + hn\} \\ \geq\ &V_{i+1}^M(t+1, n-1, m+1) + h(n-1) \\ &-\{V_{i+1}^M(t+1, n, m+2) + hn\}, \end{aligned}$$

where the last inequality follows from the assumption that $V_{i+1}^S(t+1, n-1, m) - V_{i+1}^S(t+1, n, m+1) \geq V_{i+1}^M(t+1, n-1, m) - V_{i+1}^M(t+1, n, m+1)$. Hence, we prove that

$$V_i^S(t, n-1, m) - V_i^S(t, n, m+1) \geq V_i^M(t, n-1, m) - V_i^M(t, n, m+1), \text{ when } n = 1.$$

When $n \geq 2$, we must determine whether the following inequality holds:

$$\begin{aligned}
&\min\{V_{i+1}^S(t+1, n-2, m) + h(n-2), V_{i+1}^S(t+1, n-1, m+1) + h(n-1)\} \\
&- \min\{V_{i+1}^S(t+1, n-1, m+1) + h(n-1), V_{i+1}^S(t+1, n, m+2) + hn\} \\
\geq\ &\min\{V_{i+1}^M(t+1, n-2, m) + h(n-2), V_{i+1}^M(t+1, n-1, m+1) + h(n-1)\} \\
&- \min\{V_{i+1}^M(t+1, n-1, m+1) + h(n-1), V_{i+1}^M(t+1, n, m+2) + hn\}. \qquad (6.34)
\end{aligned}$$

We prove it by checking all possible cases. We have

$$\begin{aligned}
&V_{i+1}^S(t+1, n-2, m) + h(n-2) \\
&- \min\{V_{i+1}^S(t+1, n-1, m+1) + h(n-1), V_{i+1}^S(t+1, n, m+2) + hn\} \\
\geq\ &V_{i+1}^S(t+1, n-2, m) + h(n-2) \\
&-\{V_{i+1}^S(t+1, n-1, m+1) + h(n-1)\} \\
\geq\ &V_{i+1}^M(t+1, n-2, m) + h(n-2) \\
&-\{V_{i+1}^M(t+1, n-1, m+1) + h(n-1)\} \\
\geq\ &\min\{V_{i+1}^M(t+1, n-2, m) + h(n-2), V_{i+1}^M(t+1, n-1, m+1) + h(n-1)\} \\
&-\{V_{i+1}^M(t+1, n-1, m+1) + h(n-1)\}, \qquad (6.35)
\end{aligned}$$

and

$$\begin{aligned}
&V_{i+1}^S(t+1, n-2, m) + h(n-2) \\
&- \min\{V_{i+1}^S(t+1, n-1, m+1) + h(n-1), V_{i+1}^S(t+1, n, m+2) + hn\} \\
\geq\ &V_{i+1}^S(t+1, n-2, m) + h(n-2) \\
&-\{V_{i+1}^S(t+1, n, m+2) + hn\} \\
\geq\ &V_{i+1}^M(t+1, n-2, m) + h(n-2) \\
&-\{V_{i+1}^M(t+1, n, m+2) + hn\} \\
\geq\ &\min\{V_{i+1}^M(t+1, n-2, m) + h(n-2), V_{i+1}^M(t+1, n-1, m+1) + h(n-1)\} \\
&-\{V_{i+1}^M(t+1, n, m+2) + hn\}. \qquad (6.36)
\end{aligned}$$

Based on (6.35) and (6.36), we conclude that

$$\begin{aligned}
&V_{i+1}^S(t+1, n-2, m) + h(n-2) \\
&- \min\{V_{i+1}^S(t+1, n-1, m+1) + h(n-1), V_{i+1}^S(t+1, n, m+2) + hn\} \\
\geq\ &\min\{V_{i+1}^M(t+1, n-2, m) + h(n-2), V_{i+1}^M(t+1, n-1, m+1) + h(n-1)\} \\
&- \min\{V_{i+1}^M(t+1, n-1, m+1) + h(n-1), V_{i+1}^M(t+1, n, m+2) + hn\}. \qquad (6.37)
\end{aligned}$$

Using the same approach, we can prove that

$$\begin{aligned}
&V_{i+1}^S(t+1, n-1, m+1) + h(n-1) \\
&- \min\{V_{i+1}^S(t+1, n-1, m+1) + h(n-1), V_{i+1}^S(t+1, n, m+2) + hn\} \\
\geq\ & \min\{V_{i+1}^M(t+1, n-2, m) + h(n-2), V_{i+1}^M(t+1, n-1, m+1) + h(n-1)\} \\
&- \min\{V_{i+1}^M(t+1, n-1, m+1) + h(n-1), V_{i+1}^M(t+1, n, m+2) + hn\}.
\end{aligned} \tag{6.38}$$

Then, (6.37) and (6.38) imply (6.34). Hence, we have

$$V_i^S(t, n-1, m) - V_i^S(t, n, m+1) \geq V_i^M(t, n-1, m) - V_i^M(t, n, m+1), \text{ when } n \geq 2.$$

We therefore prove that $V_i^S(t, n-1, m) - V_i^S(t, n, m+1) \geq V_i^S(t, n-1, m) - V_i^S(t, n, m+1) \geq V_i^M(t, n-1, m) - V_i^M(t, n, m+1)$. Based on this inequality, we go on to prove that $V_{i-1}^S(T, n-1, m) - V_{i-1}^S(T, n, m+1) \geq V_{i-1}^M(T, n-1, m) - V_{i-1}^M(T, n, m+1)$. From (6.9) and (6.15), when $n + Q_i \leq 0$, we know that

$$\begin{aligned}
&V_{i-1}^S(T, n-1, m) - V_{i-1}^S(T, n, m+1) \\
&= V_i^S(1, n-1+Q_i, m) - V_i^S(1, n+Q_i, m+1) \\
&= V_i^M(1, n-1+Q_i, m) - V_i^M(1, n+Q_i, m+1) \\
&= V_{i-1}^M(T, n-1, m) - V_{i-1}^M(T, n, m+1).
\end{aligned} \tag{6.39}$$

When $n + Q_i \geq 1$,

$$\begin{aligned}
&V_{i-1}^S(T, n-1, m) - V_{i-1}^S(T, n, m+1) \\
=\ & \min_{0 \leq y \leq \min\{m, n-1+Q_i\}} \{c_{pl}(m-y) + V_i^S(1, n-1+Q_i-y, 0)\} \\
&- \min_{0 \leq y \leq \min\{m+1, n+Q_i\}} \{c_{pl}(m+1-y) + V_i^S(1, n+Q_i-y, 0)\} \\
=\ & \min_{0 \leq y \leq \min\{m, n-1+Q_i\}} \{c_{pl}(m-y) + V_i^S(1, n-1+Q_i-y, 0)\} \\
&- \min_{-1 \leq y \leq \min\{m, n-1+Q_i\}} \{c_{pl}(m-y) + V_i^S(1, n-1+Q_i-y, 0)\} \\
\geq\ & 0 \\
=\ & V_{i-1}^M(T, n-1, m) - V_{i-1}^M(T, n, m+1).
\end{aligned} \tag{6.40}$$

From (6.39) and (6.40), we conclude that $V_{i-1}^S(T, n-1, m) - V_{i-1}^S(T, n, m+1) \geq V_{i-1}^M(T, n-1, m) - V_{i-1}^M(T, n, m+1)$.

Reference

X. Liang, L. Ma, H. Yan, Toyota China: Matching supply with demand, in *The Supply Chain Management Casebook: Comprehensive Coverage and Best Practices in SCM*, ed. by C. Munson (Financial Times Press, 2013), pp. 68–79

Chapter 7
Inventory Management with Alternative Delivery Times: The Case of Group Buying

Abstract In this chapter, we apply the method of modeling and analysis used in previous chapters to a discussion of the inventory control problem of group buying. The seller offers two purchasing options to its customers: a regular option and a group-buying option. The outcome of the group buying is uncertain and affects the inventory allocation at the end of the selling season. The seller thus can choose to ration the demand from regular customers to reduce possible penalty. We derive the optimal inventory rationing policy and demonstrate that it remains robust in several extensions.

7.1 Introduction

In previous chapters, we discuss inventory models with alternative delivery lead times. In those models, the seller makes use of customers' heterogeneous delivery-time requirements to segment the market by offering two alternative delivery-time options. Replenishment decisions are made in cycles, and within a replenishment cycle, the seller resorts to the inventory commitment/upgrade decision that determines the fulfillment of incoming orders to ration its limited inventory. In this chapter, we deviate slightly from this line of research and show that the same modeling and analysis approach can be applied to other problems of a similar nature. That is, when a seller must deal with multiple demand streams using a common inventory pool, it faces the same trade-off and decision-making problem as in the case of alternative delivery lead times. The group-buying mechanism shows such an example.

Group buying has emerged as a popular sales strategy in the new millennium. Also known as collective buying, it offers products and services at significantly reduced prices on the condition that a minimum number of buyers make the purchase. A common practice of firms that offer the group-buying option is to set a discounted price, a starting and a cutoff time, and a minimum number of customers required for the group formation. Participating customers get the product at the discounted price only if the final number of participants exceeds the predetermined minimum at the cutoff time. Group buying is similar to a quantity discount in the sense that the discount is contingent on a participation threshold; however, unlike the traditional

X. Liang et al., *Inventory Management with Alternative Delivery Times*, SpringerBriefs in Operations Management, DOI 10.1007/978-3-319-48635-2_7

quantity discount, group buying offers a social buying experience because the final outcome relies on the collective action of the buying group.

Given its popularity, group buying is now offered by many sellers as a complement to the traditional spot-selling option (i.e., the *mixed group-buying* strategy as defined in Jing and Xie (2011)). Although sellers can benefit from refined market segmentation and inventory pooling, the mixed group-buying strategy also adds extra complexity because sellers must now deal with multiple demand streams. Finding the optimal inventory management policy thus becomes a critical decision. Motivated by this real-world problem, we develop analytical models and analyze the corresponding optimal policies. The general setting of these models is as follows. The seller offers two purchasing options: a group-buying option and a regular (spot-selling) option. We call customers who choose these two options group-buying customers and regular customers, respectively. At the beginning of the selling season, the seller replenishes its inventory and announces the prices of the two options and the participation threshold for the group-buying option. Customers arrive sequentially. Given the uncertainties of the two demand streams and of the final outcome of the group-buying option, the seller can choose to reject the incoming orders from regular customers to reserve inventory for the group-buying demand. Based on the preceding setting, we investigate the optimal inventory management policy for the seller and generalize our results to a variety of extensions.

Our work fits into the growing line of research related to the group-buying mechanism. Previous work has explored the rationale behind the group-buying business model (Dholakia 2010; Edelman et al. 2011; Jing and Xie 2011; Hu et al. 2013; Liang et al. 2014; Wu et al. 2014) and analyzed customer behavior using empirical evidence (Kauffman and Wang 2001, 2002; Kauffman et al. 2010a, b; Zhou et al. 2013). However, less attention has been paid to the operational side of the problem. We characterize the real-time inventory control policy for a seller simultaneously dealing with two demand streams and show that this optimal policy is robust to a variety of generalizations. These results are of practical value for an effective implementation of the group-buying mechanism.

In the remainder of this chapter, Sect. 7.2 presents the formulation of the problem, and Sect. 7.3 gives the optimal inventory control policy. In Sect. 7.4, we examine several extensions of the basic model. Section 7.5 concludes the chapter.

7.2 Model Formulation

Consider a seller that offers two purchasing options to its customers over a finite time horizon: a regular option and a group-buying option. Customers choosing the regular and group-buying options are called *regular* and *group-buying* customers, respectively. The regular option, priced at p_s, delivers the product to customers immediately. The group-buying option, priced at $p_g < p_s$, backlogs participating customers and compares the number of total participating customers with a pre-announced threshold G at the cutoff time to determine whether the group buying is successful. If the final number of participants is equal to or greater than G, the group buying is successful and all participating customers get the product at p_g; otherwise, the group buying fails and all participating customers get nothing. We assume the cutoff time coincides with the end of the selling horizon.

We also divide the selling horizon into T small periods, indexed from 1 to T. In period t, the on-hand inventory level and number of backlogged participating customers are denoted as $n(t)$ and $g(t)$, respectively. Within each period before the last period, we assume that at most one customer arrives. There are three possibilities in terms of customer arrival: a regular customer, a group-buying customer, and no customer. The corresponding probabilities are denoted as π_r, π_g, and π_0, respectively. (Note that the rates can be functions of G and p_g and, in the process, be dependent on t and g to reflect customers' real-time evaluation of the group-buying option. However, these considerations do not affect our analysis approach and the results. We thus simply use π_r, π_g, and π_0 throughout the paper.) When a stock-out happens, the regular demand is lost, incurring a penalty c_p. The seller can also choose to actively reject a regular customer at a cost c_p. This demand rejection is to reserve inventory for the possible group-buying demand. At the end of the selling season, if the group buying is successful, the seller must satisfy all of the group-buying customers, and a shortfall incurs a unit penalty cost c_g, which incorporates the cost of delayed fulfillment and the marginal cost to satisfy the excess demand (e.g., the seller may have to source from other sellers to cover the shortage). The inventory holding cost is calculated in each period before the last period. Let h be the per-unit holding cost. In period T, any leftover inventory is salvaged at s. To avoid trivial cases, we assume that the following relations hold: $s < p_g < p_s$ and $p_g < c_g$.

Given the preceding model setting, we now write out the dynamic programming equations. When $1 < t \leq T$,

$$\begin{aligned} V(t-1,n,g) =& \pi_r \max\{V(t,n-1,g)+p_s-h(n-1),\, V(t,n,g)-c_p-hn\} \\ &+ \pi_g[V(t,n,g+1)-hn]+\pi_0[V(t,n,g)-hn], \quad \text{for } n>0, \end{aligned} \tag{7.1}$$

$$V(t-1,0,g) = \pi_r[V(t,0,g)-c_p]+\pi_g V(t,0,g+1)+\pi_0 V(t,0,g), \tag{7.2}$$

$$V(T,n,g) = \begin{cases} sn, & \text{if } g < G; \\ s(n-g)^+ + p_g g - c_g(g-n)^+, & \text{if } g \geq G. \end{cases}$$

The seller achieves a higher margin on regular customers and at the same time ameliorates the inventory risk by introducing the group-buying option. Through the inventory control, the seller can further reduce the risk of overselling.

7.3 Optimal Inventory Control Policy

Given that customers arrive sequentially and that the outcome of the group buying is uncertain, the critical problem for the seller is to find the optimal inventory control policy. That is, the seller must determine how to allocate the inventory between the two classes of customers in real time. We first prove the following structural property.

Lemma 7.1 *$V(t, n, g)$ is concave in n for given g.*

The concavity in Lemma 7.1 implies that the optimal inventory control policy is of a threshold type.

Theorem 7.1 *In each period t, for given $g(t)$, there exists a threshold $n^*(t)$, defined as the minimum $n(t)$ that satisfies $V(t, n-1, g) - V(t, n, g) \geq -h - p_s - c_p$, where the optimal policy is to reject the incoming regular customer when $n(t) \leq n^*(t)$ and to accept it otherwise.*

In some extreme cases, the derivation of the threshold can be greatly simplified.

Corollary 7.1 *If $p_s + c_p + hT \geq c_g$, $n^*(t) = 0$ for all t.*

The interpretation of Corollary 7.1 is clear: when the cost of delaying a group-buying customer or meeting the excess demand is lower than that of losing a regular customer, the seller tends to prefer regular customers over group-buying customers.

With the property of the threshold, we have the following result.

Theorem 7.2 *$V(t, n, g)$ is supermodular in n and g.*

Theorem 7.2 implies that in period t, $n^*(t)$ is non-decreasing in $g(t)$. That is, group-buying customers erode the quota for regular customers.

7.4 Elaborations of the Basic Model

The basic model we discuss in preceding analysis can be extended in a number of directions. In this section, we consider three such directions: batch demand, quantity ceiling, and multiple group-buying thresholds. Using an analysis approach similar to that used for the basic model, we show that most of the results we obtain for the basic model still hold without significant changes.

7.4.1 Batch Demand

We consider the case of batch demand, in which customers, both regular and group-buying, are allowed to purchase multiple units. Correspondingly, the seller now can choose to partially fulfill a batch order placed by regular customers.

Assume the purchasing amount of a regular customer is $i \in \{1, \cdots, I\}$ units with a probability $\pi_{r,i}$, and that of a group-buying customer is $j \in \{1, \cdots, J\}$ units with a probability $\pi_{g,j}$. The dynamic programming equation can be written as follows (for simple exposition, we write out only the equation when $n \geq I$).

$$
\begin{aligned}
V(t,n,g) = & \sum_{i=1}^{I} \pi_{r,i} \max_{0 \le k \le i} [V(t+1, n-k, g) + p_s k - h(n-k) - c_p(i-k)] \\
& + \sum_{j=1}^{J} \pi_{g,j} [V(t+1, n, g+j) - hn] + \pi_0 [V(t+1, n, g) - hn].
\end{aligned} \tag{7.3}
$$

We can verify that all of our preceding results remain valid in the batch-demand case.

Theorem 7.3 *$V(t, n, g)$ is concave in n for given g and supermodular in n and g.*

7.4.2 Group Buying with a Quantity Ceiling

In real-world group-buying practices, in addition to the activation threshold, a quantity ceiling (denoted by $\overline{G}$) is often imposed at the same time, which designates the maximum quantity available for the group buying. That is, the seller ceases to take in new group-buying customers if the number of participating customers reaches $\overline{G}$.

There are two different but equivalent ways to incorporate this constraint into the base model. The first way is to incorporate the effect of $\overline{G}$ during the selling horizon. When $g < \overline{G}$, the dynamic programming equations remain the same as in Eqs. (7.1) and (7.2). When $g = \overline{G}$, we have

$$
\begin{aligned}
V(t-1, n, \overline{G}) = & \pi_r \max\{V(t, n-1, \overline{G}) + p_s - h(n-1), V(t, n, \overline{G}) - c_p - hn\} \\
& + (\pi_g + \pi_0)[V(t, n, \overline{G}) - hn] \text{ for } n > 0, \\
V(t-1, 0, \overline{G}) = & \pi_r [V(t, 0, \overline{G}) - c_p] + (\pi_g + \pi_0) V(t, 0, \overline{G}).
\end{aligned}
$$

The other way is to reflect the quantity ceiling in the terminal function while leaving the inventory dynamics unchanged. In this case, the terminal function can be written as

$$
V(T, n, g) = \begin{cases}
sn, & \text{if } g < G; \\
s(n-g)^+ + p_g g - c_g (g-n)^+, & \text{if } G \le g < \overline{G}; \\
s(n-\overline{G})^+ + p_g g - c_g(\overline{G}-n)^+ & \text{if } g \ge \overline{G}. \\
-p_g(g - \overline{G}), &
\end{cases}
$$

The term $-p_g(g - \overline{G})$ can be considered as the refund to those extra group-buying customers when g exceeds $\overline{G}$. We can see that the dynamic programming equations remain the same as in the basic model, and the new terminal function also satisfies the concavity. All of the preceding results thus still hold and the optimal inventory control policy is a threshold-type rationing policy.

7.4.3 *Multiple Group-Buying Thresholds*

A more general yet more complex form of group buying allows multiple group-buying thresholds with progressive discounting. Assume that there are M thresholds, denoted by $G_1 < \cdots < G_M$. These M thresholds are linked to M oppositely ordered discounted prices, denoted by $p_{g1} > \cdots > p_{gM}$. The terminal function can be rewritten as follows:

$$V(T, n, g) = \begin{cases} sn, & \text{if } g < G_1; \\ s(n-g)^+ + p_{g1}g - c_g(g-n)^+, & \text{if } G_1 \leq g < G_2; \\ \cdots & \\ s(n-g)^+ + p_{gM}g - c_g(g-n)^+, & g \geq G_M. \end{cases}$$

The change at the terminal function does not affect the concavity, and the optimal inventory control policy remains of a threshold rationing type. However, the supermodularity no longer holds now. An intuitive explanation is that as $g(t)$ increases, the probability of reaching a higher group-buying threshold also increases. This reduces the margin the seller can get from group-buying customers, which in turn counteracts the natural tendency of increasing $n^*(t)$ as $g(t)$ increases. Hence, $n^*(t)$ is not necessarily increasing in $g(t)$.

7.5 Concluding Remarks

In this chapter, using techniques similar to those used in inventory models with alternative lead times, we discuss the inventory control problem under the mixed group-buying strategy. We show that the optimal inventory rationing policy is of a threshold type and that the rationing threshold is contingent on the prevailing number of customers who participate in the group buying. We also demonstrate that the results remain robust in several extensions.

Appendix

Proof of Lemma 7.1

Proof We use backward induction to prove the result. It is easy to verify that the property is satisfied at the terminal. Given that $V(t, n, g)$ is concave in $n(t)$, we now prove that $V(t-1, n, g)$ is also concave. That is,

$$V(t-1, n+1, g) - V(t-1, n, g) \leq V(t-1, n, g) - V(t-1, n-1, g).$$

From Eq. (7.1), when $n(t) > 0$, it is proved that the maximization operator preserves the concavity. Both $V(t, n-1, g) + p_s - h(n-1)$ and $V(t, n, g) - c_p - hn$ are concave in $n(t)$. It is clear that $V(t, n-1, g) - h(n-1)$ is merely a shift of $V(t, n, g) - hn$ and that there is at most one junction point within the feasible support that $n(t) \in [0, \infty)$. (We omit p_s and c_p as they do not affect the analysis). We must thus prove that the concavity holds at the junction point. Assuming the junction point is $n' > 0$ (the proof is straightforward when $n' = 0$), we know that the two forward differences at n' and $n' - 1$ are equal. The concavity is thus preserved by the maximization operator and $V(t, n, g)$ is concave in $n(t)$. The concavity at the boundary when $n(t) = 0$ can be proved in the same manner.

Proof of Corollary 7.1

Proof Given that $V(t, n, g)$ is concave in n, to ensure $n^*(t) = 0$, we must prove only that $V(t, 1, g) - V(t, 0, g) \leq p_s + c_p + h$ for $\forall t$. At the terminal, we know that $V(T, 1, g) - V(T, 0, g) \leq c_g$. Hence, if $c_g \leq p_s + c_p + h$, the condition is satisfied. In period $T - 1$,

$$\begin{aligned}
&V(T-1, 1, g) - V(T-1, 0, g)\\
&= \pi_r[V(T, 0, g) + p_s] + \pi_g[V(T, 1, g+1) - h] + \pi_0[V(T, 1, g) - h]\\
&\quad -\pi_r[V(T, 0, g) - c_p] + \pi_g V(T, 0, g+1) + \pi_0 V(T, 0, g)\\
&= \pi_r(p_s + c_p) + \pi_g[V(T, 1, g+1) - V(T, 0, g+1) - h]\\
&\quad +\pi_0[V(T, 1, g) - V(T, 0, g) - h]\\
&\leq \pi_r(p_s + c_p) + (\pi_g + \pi_0)(c_g - h)\\
&\leq \max(p_s + c_p, c_g - h).
\end{aligned}$$

When we repeat this procedure in period $t < T$,

$$
\begin{aligned}
V(t,1,g)-V(t,0,g) \leq\ & \pi_r \sum_{i=0}^{T-t}(\pi_g+\pi_0)^i(p_s+c_p) \\
& +(\pi_g+\pi_0)^{T-t}c_g-\sum_{j=1}^{T-t}(\pi_g+\pi_0)^j h \\
\leq\ & \max(p_s+c_p, c_g-(T-t)h).
\end{aligned}
$$

We know that

$$
\begin{aligned}
& \max(p_s+c_p, c_g-(T-t)h) \leq p_s+c_p+h \\
\iff\ & c_g \leq p_s+c_p+(T-t+1)h.
\end{aligned}
$$

We need the inequality to hold in period 1, which means $c_g \leq p_s+c_p+hT$. This completes the proof.

Proof of Theorem 7.2

Proof We use backward induction to prove the result. It is obvious that the terminal function satisfies the supermodularity. Assuming that $V(t,n,g)$ is supermodular in $n(t)$ and $g(t)$, we now prove that $V(t-1,n,g)$ is also supermodular. That is,

$$
V(t-1,n,g+1)-V(t-1,n-1,g+1) \geq V(t-1,n,g)-V(t-1,n-1,g).
$$

We first consider the case when $n(t-1)>1$. Similar to the proof of the concavity of $V(t,n,g)$ in $n(t)$, because supermodularity is preserved under addition, we must prove only that it is also preserved by the maximization operator. To be specific, we prove that

$$
\begin{aligned}
& \max\{V(t,n-1,g+1)+p_s-h(n-1), V(t,n,g+1)-c_p-hn\} \\
& -\max\{V(t,n-2,g+1)+p_s-h(n-1), V(t,n-1,g+1)-c_p-hn\} \\
\geq\ & \max\{V(t,n-1,g)+p_s-h(n-1), V(t,n,g)-c_p-hn\} \\
& -\max\{V(t,n-2,g)+p_s-h(n-1), V(t,n-1,g)-c_p-hn\}.
\end{aligned}
$$

According to the concavity and supermodularity of $V(t-1,n,g)$, the proof of the preceding inequality can be reduced to the proof of two specific cases:

$$
\begin{aligned}
& V(t,n-1,g+1)+p_s-h(n-1)-[V(t,n-1,g+1)-c_p-hn] \\
& \geq V(t,n-1,g)+p_s-h(n-1)-[V(t,n-2,g)+p_s-h(n-1)] \\
\iff & p_s+c_p+h \geq V(t,n-1,g)-V(t,n-2,g).
\end{aligned}
$$

and

$$
\begin{aligned}
& V(t,n,g+1) - c_p - hn - [V(t,n-1,g+1) - c_p - hn] \\
& \geq V(t,n-1,g) + p_s - h(n-1) - [V(t,n-1,g) - c_p - hn] \\
\Longleftrightarrow & V(t,n,g+1) - V(t,n-1,g+1) \geq p_s + c_p + h.
\end{aligned}
$$

It is clear that both inequalities hold. The supermodularity thus also holds for $V(t,n,g)$. The case where $n(t-1) = 1$ can be proved in the same manner.

Proof of Theorem 7.3

Proof We first prove the concavity by induction. It is clear that the terminal function satisfies the concavity. From Eq. (7.3), we can prove the concavity by proving that for each individual maximization operator, the concavity is preserved. The detailed proof follows the procedure as in the proof of Lemma 7.1 without significant modifications. We skip the details here.

Based on the same logic, we prove the supermodularity by proving that it is preserved by each individual maximization operator. The proof of supermodularity is thus similar to that of Theorem 7.2. We skip the details here.

References

U. Dholakia, How Effective are Groupon Promotions for Businesses? Working Paper (Rice University, 2010)

B. Edelman, S. Jaffe, S. Kominers, To groupon or not to groupon: the profitability of deep discounts. Marketing Lett. **27**(1), 39–53 (2011)

M. Hu, M. Shi, J. Wu, Simultaneous versus sequential group-buying mechanisms. Manag. Sci. **59**(12), 2805–2822 (2013)

X. Jing, J. Xie, Group buying: a new mechanism for selling through social interactions. Manag. Sci. **57**(8), 1354–1372 (2011)

R.J. Kauffman, B. Wang, New buyers' arrival under dynamic pricing market microstructure: the case of group-buying discounts on the Internet. J. Manag. Inf. Syst. **18**(2), 157–188 (2001)

R.J. Kauffman, B. Wang, Bid together, buy together: on the efficacy of group-buying business models in internet-based selling, in *The E-Business Handbook*, ed. by P.B. Lowry, J.O. Cherrington, R.R. Watson (CRC Press, Boca Raton, FL, 2002), pp. 99–137

R.J. Kauffman, H. Lai, C.T. Ho, Incentive mechanisms, fairness and participation in online group-buying auctions. Electron. Commer. Res. Appl. **9**(3), 249–262 (2010a)

R.J. Kauffman, H. Lai, H.C. Lin, Consumer adoption of group-buying auctions: an experimental study. Inf. Technol. Manag. **11**(4), 1–21 (2010b)

X. Liang, L. Ma, L. Xie, H. Yan, The informational aspect of the group-buying mechanism. Eur. J. Oper. Res. **234**(1), 331–340 (2014)

J. Wu, M. Shi, M. Hu, Threshold effects in online group buying. Manag. Sci. **61**(9), 2025–2040 (2014)

G. Zhou, K.Q. Xu, S.Y. Liao, Do starting and ending effects in fixed-price group-buying differ? an empirical analysis based on large-scale datasets from China. Electron. Commer. Res. Appl. **12**(2), 78–89 (2013)

Index

X. Liang et al., *Inventory Management with Alternative Delivery Times*,
SpringerBriefs in Operations Management, DOI 10.1007/978-3-319-48635-2

Kolibri 360 GmbH
Ferdinand-Jühlke-Straße 7
99095 Erfurt, Deutschland
produktsicherheit@kolibri360.de